Educational

Connected Minds

TECHNOLOGY AND TODAY'S LEARNERS

Francesc Pedró

This work is published on the responsibility of the Secretary-General of the OECD. The opinions expressed and arguments employed herein do not necessarily reflect the official views of the Organisation or of the governments of its member countries.

This document and any map included herein are without prejudice to the status of or sovereignty over any territory, to the delimitation of international frontiers and boundaries and to the name of any territory, city or area.

Please cite this publication as:
OECD (2012), *Connected Minds: Technology and Today's Learners*, Educational Research and Innovation, OECD Publishing.
http://dx.doi.org/10.1787/9789264111011-en

ISBN 978-92-64-07522-1 (print)
ISBN 978-92-64-11101-1 (PDF)

Series: Educational Research and Innovation
ISSN 2076-9660 (print)
ISSN 2076-9679 (online)

The statistical data for Israel are supplied by and under the responsibility of the relevant Israeli authorities. The use of such data by the OECD is without prejudice to the status of the Golan Heights, East Jerusalem and Israeli settlements in the West Bank under the terms of international law.

Note by Turkey:
The information in this document with reference to "Cyprus" relates to the southern part of the Island. There is no single authority representing both Turkish and Greek Cypriot people on the Island. Turkey recognises the Turkish Republic of Northern Cyprus (TRNC). Until a lasting and equitable solution is found within the context of the United Nations, Turkey shall preserve its position concerning the "Cyprus issue".

Note by all the European Union Member States of the OECD and the European Commission:
The Republic of Cyprus is recognised by all members of the United Nations with the exception of Turkey. The information in this document relates to the area under the effective control of the Government of the Republic of Cyprus.

Photo credits: Cover © Eric Audras/Onoky/Corbis

Corrigenda to OECD publications may be found on line at: *www.oecd.org/publishing/corrigenda*.

Foreword

Back in 2007, when the OECD work on the New Millennium Learners (NML) started, there was a sense that the increasing levels of technology attachment of young people, their familiarity with digital media and the fact that they are always connected, had to have, sooner or later, an impact on education. This was the intuition that led the OECD Centre for Educational Research and Innovation (CERI) to tackle the issue from an evidence-based perspective. As the ongoing discussions about the New Millennium Learners tend to be pervaded by previous assumptions about learning, on the one hand, and technology, on the other, it was felt that an evidence-based perspective is the only one that can bring answers to the many issues prompted, paving the way for appropriate policy responses.

In particular, CERI's work in this area addressed three broad questions. First, can the claim that today's students are New Millennium Learners, or digital natives, be sustained empirically? Second, is there consistent research evidence demonstrating the effects of technology adoption on cognitive development, social values and learning expectations? Third, what are the implications for educational policy and practice?

With this report CERI provides a coherent and comprehensive set of answers to these research questions, with which the OECD wants to contribute to the debate about the effects of technology attachment and connectedness on learners, particularly in relation to their expectations regarding teaching.

This is the final report of CERI's major project in the area of digital media and education, under the general title of the New Millennium Learners (NML). This project was launched in 2007 and its main goal has been to analyse the new generation of learners and to understand their expectations and attitudes towards education, drawing on empirical evidence. The impact of digital technologies on cognitive skills and on learning expectations, and the evolution of social values and lifestyles, as well as the relationships between technology use and educational performance have been the most important issues addressed.

Similarly, the evidence shows that young learners' expectations and behaviours in relation to technology use or connectivity in formal education are not changing dramatically – at least not yet. Rather, students tend to be more reluctant in this respect than the image of the new millennium learner might suggest. Most of them do not want technology to bring a radical transformation in teaching and learning but would like to benefit more from their added convenience and increased productivity gains in academic work. If those gains do not become apparent to students, then reluctance emerges. The reasons for such reluctance might be related to the uncertainty, disruptiveness and discomfort that discrete technology-based innovations not clearly leading to learning improvements may cause to them.

The report elaborates also on the policy responses. It acknowledges that governments have done a lot to support technology adoption in teaching, with important investments in infrastructure as well as in services both for students and teachers, and educational institutions and their networks as well. A first key message is that governments must keep up with emerging technology developments, equipment and applications, and contribute to supporting innovations intended to explore the value and possible benefits of technology adoption for teaching and learning.

Because of the growing importance of connectedness, schools and teachers must cope with new responsibilities related in particular to skills with which they may not be as familiar as necessary. Yet, an increasing percentage of teachers can hardly be considered digital immigrants. The adoption of technology has contributed to transforming teachers' work although this process is slower in the schools sector than in higher education. There are indications that the actual use of technology in teaching in higher education clearly outperforms the equivalent in the schools sector in most OECD countries. But the gap in technology adoption between students and teachers is decreasing although the range of applications and services used by them differ. The key message here is that neither schools nor teachers can be said to be closing their eyes to changes in students' behaviours, needs and expectations. But the responsiveness of education to them could be quicker.

The work on the New Millennium Learners has received direct contributions from a number of countries, namely those of Austria, Chile, Italy, Norway, South Korea and the United States, as well as from the Knowledge Foundation (Stockholm), the Jaume Bofill Foundation (Barcelona). We are also very grateful to the John D. and Catherine T. MacArthur Foundation (Chicago) for their generous support.

Francesc Pedró, formerly at CERI and currently at UNESCO, managed the drafting of this report, with significant contributions from Oscar Valiente (Chapters 1 and 2), Katerina Ananiadou (Chapters 2 and 4) and Emanuele Rapetti (Chapters 3 and 6). At various iterations, different chapters of the manuscript received invaluable feedback from Prof. Georges-Louis Baron (France), Doug Brown (England), Oystein Johannessen (Norway) and Yngve Wallin (Sweden). The final version benefitted from various comments and suggestions of the members of the CERI Governing Board. This work would not have been possible without the assistance of Ashley Allen, Cassandra Davis, Stephanie Villalobos Gonzalez and Therese Walsh, as well as the editorial work of Lynda Hawe and Alys Barber. Also, we would also like to thank Peter Vogelpoel for his work in formatting this publication.

Barbara Ischinger

Director of Education, OECD

Table of contents

Figures

Tables

Boxes

Executive summary

In all OECD countries digital media and connectedness are integral to the lives of today's learners. Indeed, it is often claimed that today's students are "New Millennium Learners or "Digital Natives" and have different expectations about education than previous generations. In many OECD countries this is not so surprising as it applies also to a growing percentage of the adult population.

The debates about the implications of this phenomenon for education have been already taking place for some time. Powerful and suggestive images, like the "digital natives", have emerged to evoke and summarise in an intuitive form a given set of expectations about today's learners. Whether or not the level of technology adoption or dependence is having an impact on the way students manage knowledge and learn and, therefore, on their expectations about teaching and learning has been open to discussion. Such a discussion has often devolved into an irreconcilable confrontation between the advocates of educational change and those who look at technology in teaching merely as a tool to perfect what teachers do. The former see in the new generations of technology-adept students an opportunity to radically transform teaching and learning in formal education. The latter claim that technology should be used to enhance and improve current practices and that the level of technology adoption should be a function of two criteria: convenience and productivity.

Nevertheless, rarely are these discussions backed by empirical evidence. Many works on this topic can be seen as stimulating and challenging essays that expand the horizon of the debates and drive the debates by anticipating plausible hypotheses. However, they often fail to provide the empirical evidence that could contribute to informing the policy debate at institutional level and to help teachers to make professional choices about technology adoption in teaching on a sound basis.

A first key message is that governments must keep up with emerging technology developments, equipment and applications, and contribute to supporting innovations intended to explore the value and possible benefits of technology adoption for teaching and learning.

Three main questions are addressed here in order to contribute to filling this knowledge gap. First, can the claim that today's students are New Millennium Learners, or digital natives, be sustained empirically? Second, is there consistent research evidence demonstrating the effects of technology adoption on cognitive development, social values, and learning expectations? Third, what are the implications for educational policy and practice?

The responses found suggest a mixed and far more complex picture than is usually presented in most of the well-known essays on this topic. To begin with, although an increasing percentage of young people can be said to be adept in technology, it is misleading to assume that all of them fit equally well into the image of New Millennium Learners. As is the case with learning styles, there are different student profiles regarding technology adoption and use, and in many respects clear digital divides still exist. The use of concepts such as the New Millennium Learners or digital natives can be helpful in so far as such concepts evoke a clear and powerful image but misleading if they are used as clichés or stereotypes. For the purposes of improving teaching and learning in formal education, it is the diversity of students' practices, preferences and situations in relation to technology that matters most.

Secondly, there is not enough empirical evidence yet to support the idea that students' use of technology and digital media is transforming the way in which they learn, their social values and lifestyles, and, finally, their expectations about teaching and learning. In particular, students' attitudes towards technology use in teaching and learning appear to be far from what many would wish to be the dominant patterns that would emerge. Rather, students tend to be more reluctant in this respect than the image of the New Millennium Learner might suggest. Most of them do not want technology to bring a radical transformation in teaching and learning but would like to benefit more from their added convenience and increased productivity gains in academic work. If those gains do not become apparent to students, then reluctance emerges. The reasons for such reluctance might be related to the uncertainty, disruptiveness and discomfort that discrete technology-based innovations, which do not clearly lead, to learning improvements may cause to them.

The report clearly demonstrates that, as of today, there is not enough research evidence to demonstrate that technology attachment or connectedness has critical effects on cognitive skills development. It may be too early to perceive significant effects, however, there are some indications that in the long run, due to continued practice, and verbal intelligence levels may decrease to the benefit of image or spatial intelligence. Yet, claims about changes in the brain caused by attachment to technology or connectedness are simply not backed by evidence.

While educational institutions and teachers are increasingly adopting technology in teaching, there is an urgent need to address this issue in a systemic way. This means identifying which policies and practices will best suit the objective of providing students with a rich learning environment while improving their satisfaction, with convincing reasons based on effective practice, and thus boosting learning gains. More must be done to improve the knowledge base about technology use in education so as to inform the debates. In particular, activities intended to train and support teachers for course adoption of technology should be based on validated effective practices. All this requires not only more experimental research but also increased efforts to better disseminate existing findings and thus avoid reinventing the wheel.

Introduction

Why connectedness matters

Today's discussions about the impact of technology on the economy and society have to take into account the growing importance of connectivity. Connectedness, which is the capacity to benefit from connectivity for personal, social, work or economic purposes, is having an impact on all spheres of human activity. Therefore, devices and gadgets are less important than the ability to be connected and seizing the opportunities that connectedness offers. This introduction defines what connectedness is, and explains why today's policy discussions about the knowledge economy and society must shift from purely technology-related issues to the opportunities brought about by connectedness and digital media. To turn connectivity into connectedness dedicated policies have to be designed with a twofold goal: first, to guarantee that all the emerging opportunities brought about by technology and its outcomes can be seized in favour of economic and societal development and second, that the resulting benefits of these opportunities are equally accessible to all. Education has to play a major role in the achievement of these two goals. To begin with, both policy makers and educators are increasingly concerned about the fact that children and adolescents in particular are extremely attached to digital media and spend a lot of time connected. Yet they need to know how far the claims, both positive and negative, usually made about the effects of technology attachment are based on real and trustworthy evidence; secondly, what courses of action may make sense in the context of the emergence of knowledge economy and society; and finally, how attachment to digital media and connectivity during childhood and adolescence can be used in favour of promoting connectedness.

What individuals, social groups, institutions, firms or governments can do is evolving rapidly due to connectivity, that is their ability to link with others, be that through dedicated networks, fixed or mobile, or through the Internet. Examples of the applications of connectivity can be found in almost every domain of human activity. Connectivity can be seen as a gateway to both transforming existing processes and creating new ones in particular in relation to:

- **Information**, from production to access, from accumulation to sharing.
- **Services**, be they commercial or free, public or private.
- **People**, so as to reinforce existing social networks or to be incorporated into new ones.

Technological change has made the state of being connected, or not, far more important today for the economy, for society and for individuals than the vast array of technologies, devices and gadgets that can grant connectivity. Not surprisingly, the debates about whether connectivity is a public good and the ability of being connected has to be considered a right, and therefore granted by law, or not, are increasingly shaping the policy agenda internationally.

It's not about technology, it's about connectedness

There are different meanings attached to the broad concept of information and communication technologies (ICT). In the early eighties the concept, sometimes restricted to information technology (IT) referred almost exclusively to computers and their early applications, namely word processing, database management and calculation. The concept grew later on to also cover other devices intended to digitally support media, such as laser discs and DVDs. With the emergence of the Internet in the mid-nineties, the concept expanded to encompass all technologies and applications intended to support communication and provide access to digital information and media, such as those embodied in broadband infrastructures or in e-mail and Internet browser applications. Finally, mobile phones with access to the Internet contributed to expanding the concept of ICT so that it has now reached its current broader and all-encompassing meaning. Therefore, generally speaking, the expression "information and communication technologies" comprises all these elements, namely computers, networks, mobile phones and all the hybrids or new technology developments such as smartphones, tablets, digital pads or netbooks as well as the applications that can be run on them.

This is not without problems, however. Although the concept of technology or ICT was a useful construct in the eighties and in the nineties, since the progressive generalisation of access to the Internet, what really matters is the ability to stay connected either to others or to the Internet, irrespective of the type of device, service or platform used. This consideration is important because it helps to focus the policy discussion in education, as in other sectors, not on issues of access to particular types of technology, devices or gadgets, as was the case with the issue of computers in the classroom back in the nineties, but rather on the vast range of activities that can be carried out or services that can be accessed while being connected.

Being connected does not necessarily translate into immediate benefits. This is why a new concept is proposed here, that of connectedness. Connectedness can be defined as the capacity to benefit from connectivity for personal, social, work or economic purposes. To turn connectivity into connectedness and thus seize the opportunities raised by pervasive connectivity, individuals have to be capable of actively dealing with digital information and media. They also need to be able to communicate effectively with others using online digital applications or services.

Connectedness can also be described as a dual state: people, institutions, firms or governments are either connected (on) or not (off). Institutions, firms and governments should be permanently on in order to seize the opportunities of the knowledge economy and society. As the next chapter shows, there is clear evidence that an increasing percentage of citizens in OECD countries are almost permanently on, either for work or social purposes, or for both.

The marriage between digital media and connectivity is having an important effect on contemporary life. To begin with, this combination is fuelling economic growth. New business opportunities are emerging thanks to technology and, increasingly, to connectivity. For private firms, institutions and public organisations to turn connectivity into connectedness demands not only important organisational changes but also changes in the nature of work and the skills required. Secondly, the way in which people access services or goods, both public and private, is increasingly permeated by connectivity. Last but not least, social relationships are affected by it. In sum, connectivity is resulting in important societal and economic changes but granting access only does not translate opportunities into actual benefits, does not transform connectivity into connectedness.

Reasons for increased policy relevance

The progressive universalisation of connectivity has dramatically changed the way people can access information, publish their own and share it, as well as the way they can communicate with other people, anytime and anywhere, and create and nurture communities. This has enormous implications for all spheres of adult life, from work to leisure, from private business to public services – including education.

These implications require a policy response. This response should have a twofold goal: firstly, to guarantee that all the emerging opportunities brought about by technology and its outcomes, such as the ability to be connected, can be seized in favour of economic and societal development and, secondly, that the resulting benefits of these opportunities are equally accessible to all.

Education is expected to play an important role in this transformation also from a twofold perspective. On the one hand, education can equip individuals with the required skills for harnessing the opportunities that the knowledge economy and society offers. Whether the knowledge economy requires different types of skills, such as those represented by the construct of the so-called 21st century skills, or an improved and enhanced mastery of traditional competencies to better fit the increasingly important role that science, maths and technology play in the economy nowadays – or both at the same time – is subject to debate. In addition, education can assist individuals striving with identity formation and the new societal challenges that a society permeated by technology poses to all – from issues of privacy to property rights, from active participation in online communities to effective and productive knowledge management. On the other hand, education has an important role in relation to equity as well, as a recent OECD report on the NML has helped to unveil (OECD, 2010a). As the report states, schools in particular have been a useful tool so far in the struggle against the first digital gap, that of access, but in OECD countries the more subtle challenge of the second digital divide, that of use, is still open and requires the implementation of innovative compensatory policies.

There is, however, a second source of concern for educators and policy makers other than the impact of connectivity on today's world. This is the fact that children and young people are the population groups exhibiting the highest intensity of attachment to digital media, most of the time thanks to connectivity. Many voices claim that this is a huge generational change with unprecedented implications for education, and that education is precisely the lever to transform the opportunities brought about by connectivity into connectedness. Some see in it an opportunity for educational change, with young people being its most important driver. Opposite views support the idea that attachment to digital media and connectivity has far more negative effects on the education of young people than positive outcomes.

Research about technology attachment and student outcomes is severely limited by a number of methodological issue (OECD, 2010a). Yet, correlational studies seem to show a clear negative relationship between levels of media use and grades: the higher the use of media, the lower the grades. The problem with these correlational studies is they do not establish whether there is a cause and effect relationship between media use and grades. If there is such a relationship, it could well run in both directions simultaneously (Pedró, 2012). For instance, an American survey (Kaiser Family Foundation, 2010) found that children who are heavy media users are more likely to report getting fair or poor grades than other children. Indeed, nearly half (47%) of all heavy media users say they usually get fair or poor grades, compared to 23% of light media users. Moreover, the relationship between media exposure and grades withstands controls for other possibly relevant factors such as age, gender, race, parent education, personal contentedness, and single vs. two-parent households. Interestingly, the pattern varies only for print; those with lower grades spend less time reading for pleasure than other children do.

Both policy makers and educators might be wondering whether this deserves a response from their side and the best way in which technology-supported innovations can lead to systemic change in education, as another OECD report on the NML has also highlighted (OECD, 2010b). They may find concepts such as the NML or digital natives useful when describing the magnitude of the generational changes they have to face in relation to the growing importance of digital media and connectivity in young people's lives. Yet they need to know how far the claims usually made, both positive and negative, are based on real and trustworthy evidence;[1] secondly, what kinds of courses of action may make sense in the context of the emergence of knowledge economies and societies; and finally, how attachment to digital media and connectivity during childhood and adolescence can be used to promote connectedness, that is, the ability to make the most out of the pervasiveness of connectivity and digital media for individual, social and economic purposes. This is precisely what this report is about.

How this report is organised

In addition to this introduction, this report is organised in eight different chapters. The chapters can be seen as addressing three different sets of questions: making the case for the NML; contrasting hypotheses with empirical data; and drawing implications for policy, practice and research.

The first chapter addresses the extent of the impact of digital media and connectivity on the economy and society, and how this impact can translate into opportunities for increased connectedness. Chapter 2 presents some relevant indicators about the extent to which young people in particular are attached to digital media and connected, thus making eventually the case for the NML. The third chapter presents alternative views about the NML, optimistic, pessimistic and sceptical, and suggests what a sensible evidence-supported approach could look like. Chapters 4 and 5 document the evidence about the effects of digital media and connectivity, and in particular on their expectations about learning. Drawing on this, Chapter 6 redefines what the challenges posed by NML to education are, and how they can also be seen as opportunities to promote connectedness. Finally, the last chapters summarise the key findings, suggest some implications for policy makers and educators, and elaborate on the conclusions.

Note

1. A quick Google search using the term "digital native" in March 2009 (Helsper and Eynon, 2010) provided 17 400 hits worldwide for websites created in the last year and a Nexis search threw up 114 English language newspaper articles that used this term in the last year. In comparison, Web of Science only cited two and Scopus only 12 academic articles which ever mentioned this term. This suggests that while the term is popular, there is not much academic research in this area.

References

Helsper, E. and R. Eynon (2010), "Digital natives: where is the evidence?", *British Educational Research Journal*, Vol. 36(3), pp. 502-520.

Kaiser Family Foundation (2010), *Generation M2: Media in the Lives of 8-Year-Olds*, *www.kff.org/entmedia/upload/8010.pdf.*

Organisation for Economic Co-operation and Development (OECD) (2010a), *Are the New Millennium Learners Making the Grade? Technology Use and Educational Performance in PISA*, OECD Publishing, Paris.

OECD. (2010b), *Inspired by technology, driven by pedagogy: A Systemic Approach to Technology-Based School Innovations*, Paris, OECD Publishing.

Pedró, F. (2012), "Trusting the Unknown: The Effects of Technology Use in Education", in S. Dutta and B. Bilbao-Osorio (eds.), *Global Information Technology Report, 2012: Living in a hyperconnected world,* pp. 135, *http://www3.weforum.org/docs/Global_IT_Report_2012.pdf.*

Chapter 1

How connectedness is shaping the economy and society

Technology and connectivity are having an important effect on contemporary life. This chapter provides the evidence of such an effect. To begin with, technology is fuelling economic growth. New business opportunities emerge thanks to technology. The adoption of technology changes not only the way in which work is organised but also the nature of work and the skills required. Secondly, the way in which people access services or goods, both public and private, is also permeated by technology and connectivity. Last but not least, social relationships are also affected by new technology developments. Yet, the impressive effects of technology and connectivity do not immediately translate into connectedness as a public good. There are clear indications that important divides persist, particularly drawing on gender, age and socio-economic status.

Note by Turkey: "The information in this document with reference to 'Cyprus' relates to the southern part of the Island. There is no single authority representing both Turkish and Greek Cypriot people on the Island. Turkey recognises the Turkish Republic of Northern Cyprus (TRNC). Until a lasting and equitable solution is found within the context of United Nations, Turkey shall preserve its position concerning the 'Cyprus issue'."

Note by all the European Union Member States of the OECD and the European Commission: "The Republic of Cyprus is recognised by all members of the United Nations with the exception of Turkey. The information in this document relates to the area under the effective control of the Government of the Republic of Cyprus."

The statistical data for Israel are supplied by and under the responsibility of the relevant Israeli authorities. The use of such data by the OECD is without prejudice to the status of the Golan Heights, East Jerusalem and Israeli settlements in the West Bank under the terms of international law.

There has been a dramatic increase in the pervasiveness of connectivity everywhere. In just a few years, mobile phone subscriptions have outnumbered landlines and access to Internet is becoming almost universal in the OECD countries. As Figure 1.1 shows, the recent evolution (2003-08) of the number of mobile phones subscribers is even more impressive than the equivalent development of access to the Internet.

Figure 1.1. **Internet users and mobile subscriptions per 100 habitants in OECD countries, 2003-08**

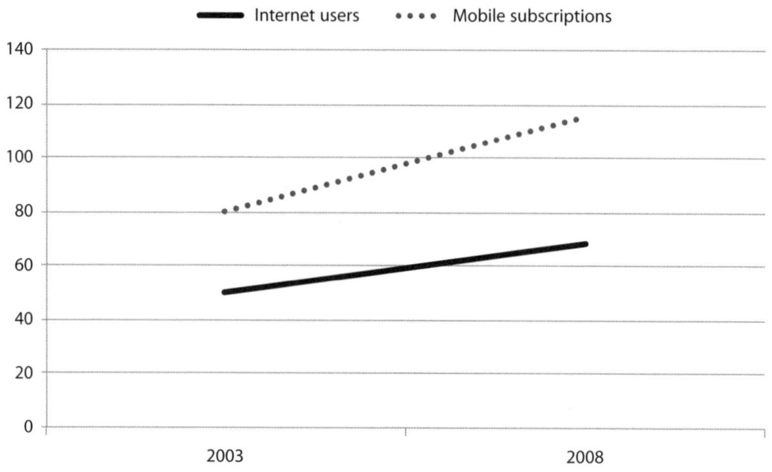

Source: ITU, Information Society Statistical Profiles, 2009.

In fact, already in 2008 there were more mobile phone subscriptions than inhabitants in most OECD countries with the only exceptions of Canada, France, Japan, Korea, Turkey and the United States. In Italy, for instance, there are 1.5 mobile phone subscriptions per inhabitant, while they were only 0.9 back in 2003. Access to the Internet has also increased substantially in the past five years, although the figures are not so impressive. In roughly one third of the OECD countries at least 75% of the population have access to the Internet, the exceptions being Greece, Italy, Poland, Portugal, and Turkey where less of half the population have access to the Internet. On the whole, however, the number of people with access to the Internet has increased by 50% in just five years and it can be expected to rise as much in the following years, thus leading to near-universal access.

The emergence and progressive universalisation of Internet access have dramatically changed the way in which people can access information, share it or publish their own, as well as their ability communicate with other people, anytime and anywhere, and create and participate in virtual communities.

This is having enormous implications in all spheres of adult life, from work to leisure, from private business to public services – including education.

Overall economic impact on growth and productivity

Over the period 1985-2006 the contribution of ICT to annual economic growth was outstanding. In the majority of OECD countries ICT investments were more important for growth than non-ICT investments. This was particularly noticeable in Australia, Belgium, Denmark, Sweden, the United Kingdom and the United States. When comparing the period 1995-2003 to that of 1990-95, the contributions of ICT investments to GDP growth accelerated in most OECD countries. The acceleration over these years was particularly significant in Australia, Belgium, France, Ireland and the United States.

Figure 1.2. **Contributions of ICT investment to GDP growth, 1990-2003, in percentage points**

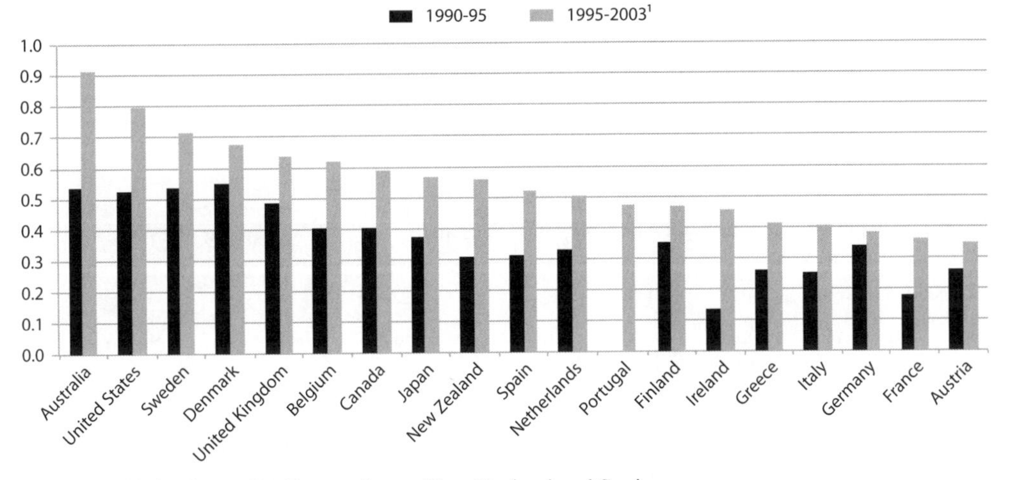

1. 1995-2002 for Australia, France, Japan, New Zealand and Spain.

Source: OECD Productivity Database, September 2005 [*www.oecd.org/statistics/productivity*].

In most OECD countries businesses have rapidly adopted the Internet. Already in 2007 approximately 95% of medium and large businesses used the Internet and 79% of them with a broadband connection. In Austria, Denmark, Finland, Iceland, Japan and Switzerland, over 98% of businesses with 10 or more employees used the Internet, and in Switzerland, Iceland, Korea, Australia, Canada and Finland over 90% of them had a broadband connection.

Figure 1.3. **Business use of broadband, 2003-10 or latest available year**

Percentage of businesses with 10 or more employees

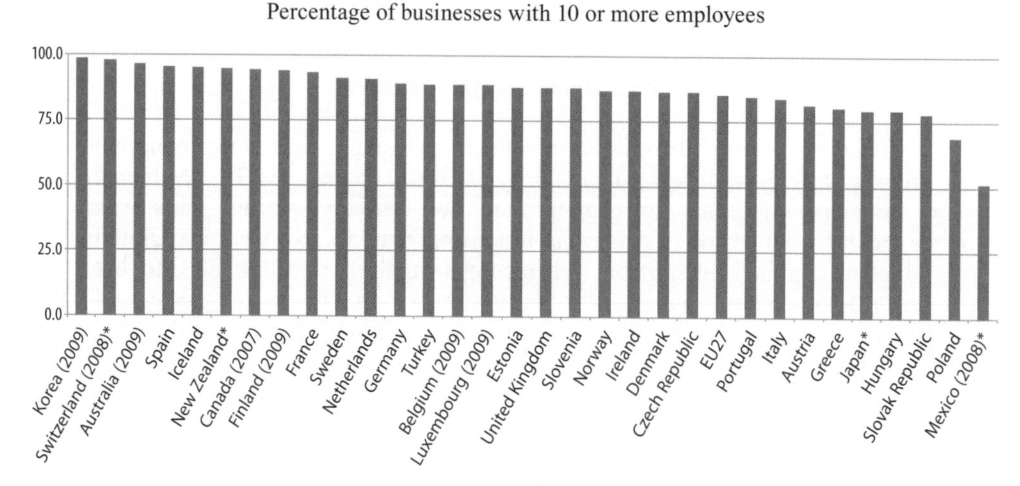

Notes:

Break in series in 2010 for EU countries: 2003-09 data based on NACE Rev 1.1, 2010 based on NACE Rev. 2, and 2003-09 data for enterprises with fixed broadband access, for 2010 data are for enterprises with fixed or mobile broadband access. Non-EU data based on fixed or wireless broadband.

For Japan, businesses with 100 or more employees.

For Mexico, businesses with 20 or more employees.

For New Zealand, businesses with 6 or more employees and with a turnover greater than NZD 30 000.

For Switzerland, businesses with 5 or more employees and connections equal to or faster than 144 Kilobits per second (mobile and fix).

Source: OECD, ICT database and Eurostat, Community Survey on ICT usage in enterprises, November 2011. Asterisk data refer only to the services sector.

A good indication of the changes and opportunities brought in by this development is, for instance, the emergence of business websites. Three out of four businesses with 10 or more employees in OECD countries had their own website in 2007 but the percentage in Japan, Sweden, Denmark, Switzerland and Finland was higher than 90%. More importantly, a growing number of businesses purchase and sell goods and services via the Internet. In 2007, on average, one third of all businesses with 10 or more employees used the Internet for purchasing and 17% for selling goods or services. In 2008, e-commerce was above 20% of total turnover in Norway, Denmark and the United Kingdom, and above 15% in Ireland and Finland. In some OECD countries, e-commerce has increased by five to seven times from the late 1990s to the mid-2000s and it is expected to grow even more in the near future.

Figure 1.4. **Firms' turnover from e-commerce, 2010**

As a percentage of total firm turnover

25.0
20.0
15.0
10.0
5.0
0.0
Ireland
Norway
Czech Republic
Iceland
Sweden
Germany
Finland
Belgium
Denmark
United Kingdom
Hungary
Netherlands
EU27
France
Austria
Portugal
Spain
Slovak Republic
Slovenia
Estonia
Switzerland (2008)
Poland
Italy
Australia (2009)
Mexico (2008)

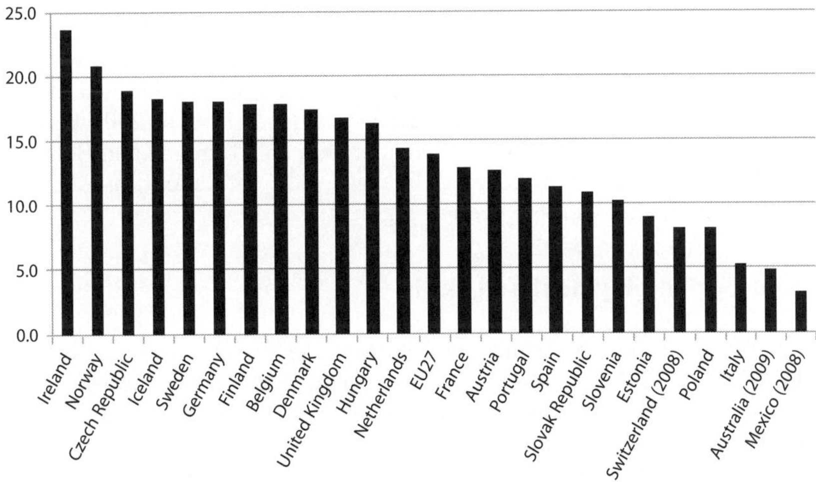

Source: ICT Database, May 2011 and Eurostat, Community Surveys on ICT Usage in Enterprises, April 2011.

Figure 1.5. **Contribution of ICT capital growth to labour productivity growth in market services**

1995-2004

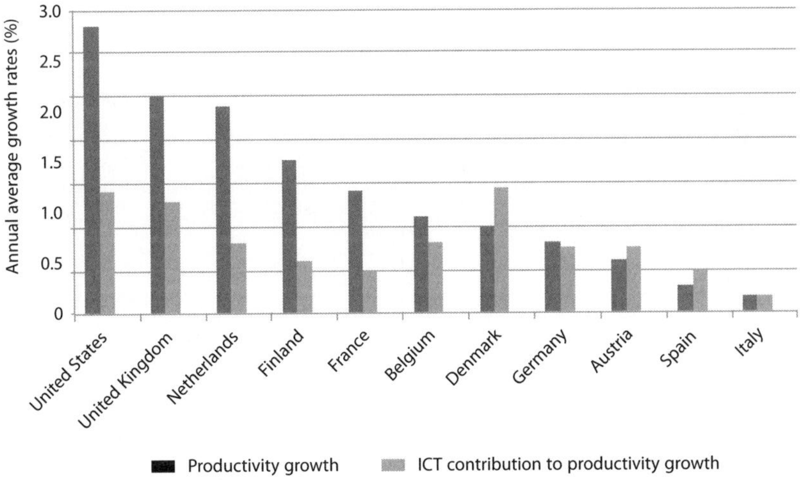

Annual average growth rates (%)
3.0
2.5
2.0
1.5
1.0
0.5
0
United States
United Kingdom
Netherlands
Finland
France
Belgium
Denmark
Germany
Austria
Spain
Italy
Productivity growth
ICT contribution to productivity growth

Source: OECD, based on The EUKLEMS productivity report, March 2007.

It is not only about investments and trade, it is also about productivity in the workplace which has also been boosted in a number of OECD countries. According to a series of studies in OECD countries, market services are the main source of overall productivity growth and technology investments are fostering relevant productivity gains in market services. For instance, in the United States, the United Kingdom and Denmark labour productivity in market services increased by over 1.2% a year from 1995 to 2004 because of technology investments. In the particular cases of Germany and Belgium, technology investments accounted for over 80% of labour productivity growth in market services, as seen in Figure 1.5.

Implications for employment and skills

Technology has also grown as a business sector over the past decade. From 1995 to 2008, growth in gross value added was higher in the information and communication technologies sector (76%) than in any other business sector. Value added in the ICT sector increased as a share of business sector value added in most OECD countries over the period 1995-2008. The largest shares were in Finland, Ireland and Korea (all over

Figure 1.6. **Share of ICT value added in business sector value added**

1995 and 2008

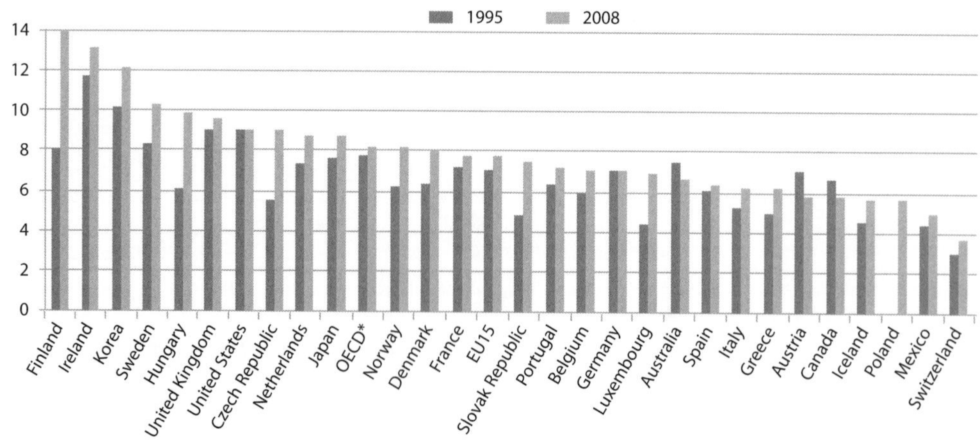

* Data for Turkey not available.

Notes: Iceland and Switzerland are 1997, New Zealand 1996. Hungary, Sweden and the United Kingdom are 2007, Canada and Portugal 2006. For Mexico latest available data are 2004.

EU15 comprised the following 15 countries: Austria, Belgium, Denmark, Finland, France, Germany, Greece, Ireland, Italy, Luxembourg, Netherlands, Portugal, Spain, Sweden, and the United Kingdom.

Source: OECD estimates, based on national sources, STAN and National Accounts databases, February 2010.

Figure 1.7. **Business R&D in the manufacturing sector by technological intensity, 2008**

As a percentage of manufacturing business enterprise R&D

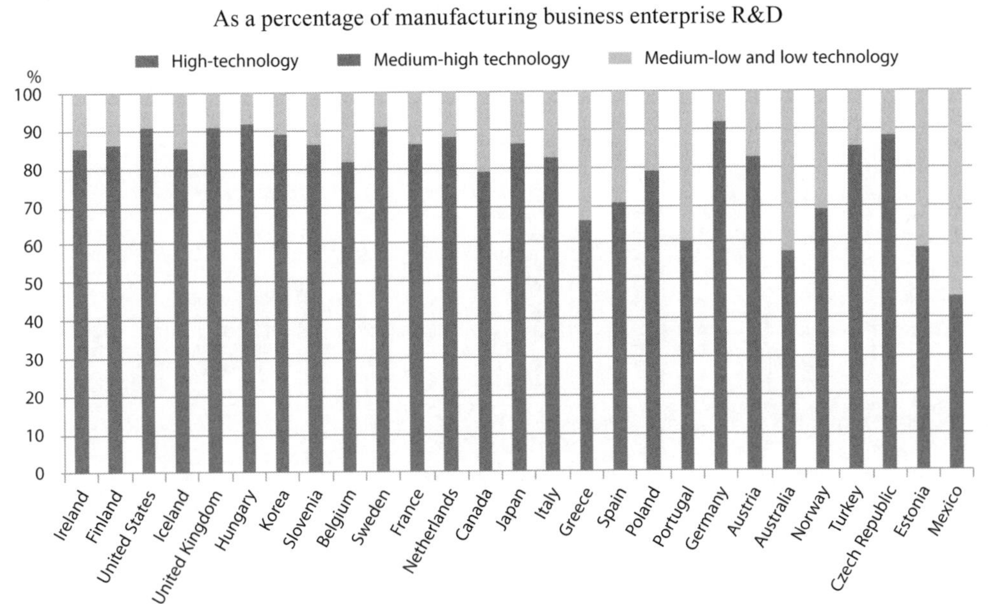

Notes:

Technology groupings give a broad sense of the relative specialisation of countries in terms of business R&D, but do not take into account the fact that in some countries the technology intensity of a given industry may be significantly different from that of the OECD average. Further details on the technology classification are available at: *http://oecd.org/dataoecd/43/41/48350231.pdf.*

2010 data for Italy. 2009 data for the Czech Republic, Estonia and Japan. 2007 data for Austria, Belgium, Canada, Finland, France, Germany, Greece, Mexico, Sweden and the United States. 2006 data for the Netherlands and Poland. 2005 data for Iceland and Ireland.

Source: OECD, ANBERD Database, May 2011.

12%) and the smallest in Switzerland, Mexico and Poland. Increasing shares were most notable in Finland, the Slovak Republic, Hungary and the Czech Republic, as well as in Sweden and Korea.

In the knowledge economy, technology stimulates innovation. Not surprisingly, the technology sector invests very heavily in research and development (R&D). In 2006, OECD area high-technology industries accounted for more than 52% of total manufacturing R&D. They accounted for over 67% of total manufacturing R&D in the United States and for 45% and 42% in the European Union and Japan, respectively. Manufacturing R&D expenditure is highly skewed towards high-technology industries in Finland, Hungary, Ireland and the United States.

Figure 1.8. **Share of high and medium-high technologies in manufacturing exports**

2007

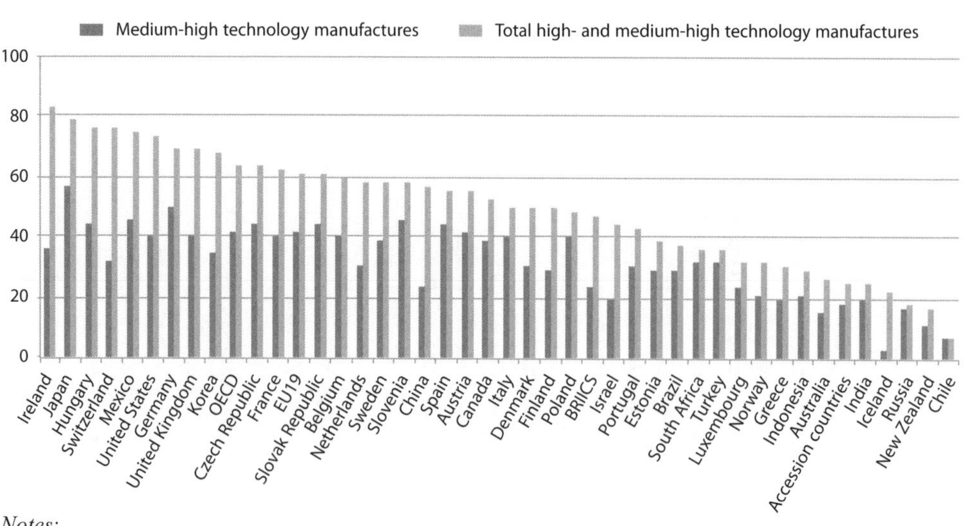

Notes:

The OECD and EU aggregates exclude Luxembourg for which data are only available from 1999.

Underlying data for China include exports to Hong Kong (China).

Source: OECD, STAN Indicators Database, 2009 edition. Underlying series from STAN Bilateral Trade Database.

Figure 1.9. **Index of the OECD trade in ICT goods and communications equipments**

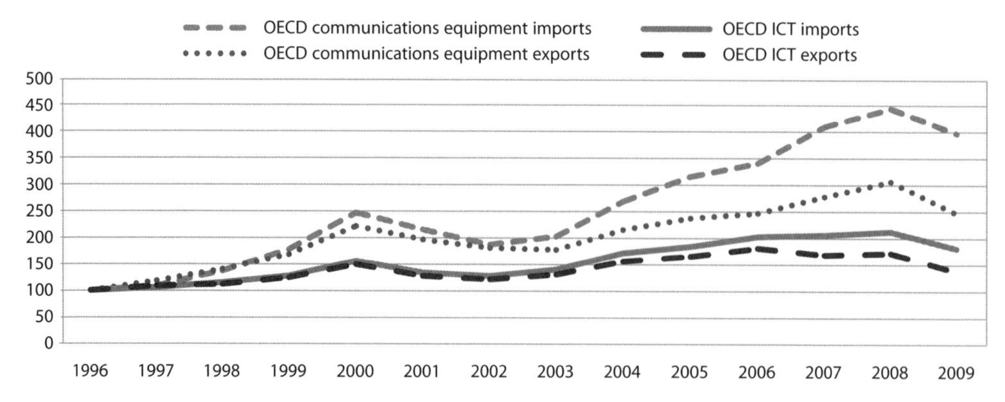

Source: OECD Communications Outlook 2011.

High-technology goods have been among the most dynamic components of international trade over the last decade and, therefore, a major determinant of countries' overall competitiveness in the world economy. Figure 1.8 shows how, in 2007, exports were particularly oriented towards high- and medium-high technology manufactures in Ireland, Japan, Hungary, Switzerland, Mexico and the United States.

In 2005, OECD member countries' exports of computer and information services totalled USD 120 billion and accounted for 3.3% of total service exports. As shown in Figure 1.9, computer and information services have been the top-ranking category in terms of growth in trade in services.

The growth of the ICT sector was driven mostly by technology services. ICT services account for more than two-thirds of total ICT sector value added in most countries and their share has grown. Overall, computer and related services and other ICT services have grown most rapidly, and more rapidly than total business services. More importantly, technology services were also the leading component of employment growth in the business sector. As shown in Figure 1.10, employment in technology services grew by 28% while in the remaining business services growth was only 20%. In the particular domain of computer and related services, employment growth was over 55% during the decade.

Figure 1.10. **Employment growth by sector in OECD countries**

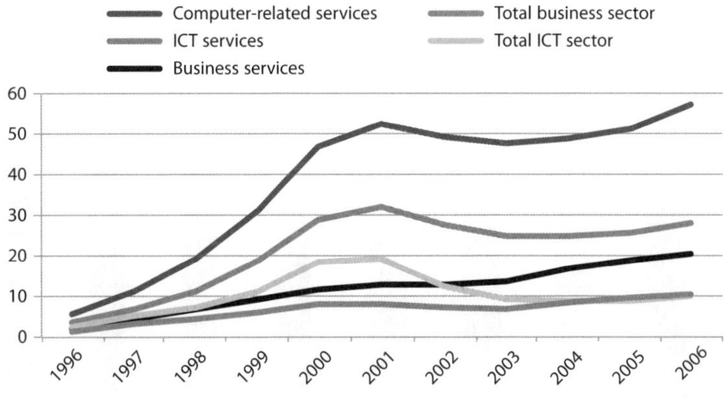

Note: The employment series for Australia starts in 1998.

Source: OECD Information Technology Outlook 2008, based on STAN database.

Almost 16 million people were employed in the ICT sector in OECD countries in 2008, 5.8% of total OECD business sector employment. Long-term growth (1995-2008) was more than 1.2% a year, almost a half percentage

Figure 1.11. **Share (%) of ICT employment in business sector employment**

1995 and 2008

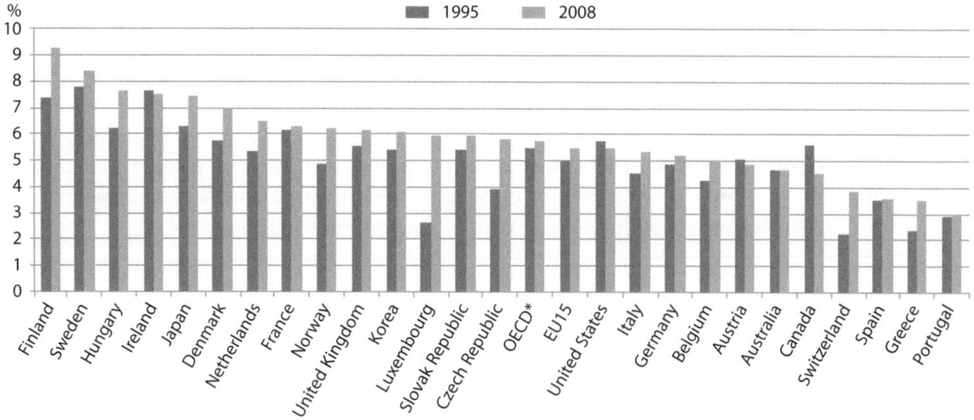

* Data for Iceland, Mexico, New Zealand, Poland and Turkey are not available.

Note: For Hungary, Portugal, Switzerland and the United States, 2007 instead of 2008. For Hungary, 2000 instead of 1995.

Source: OECD estimates, based on national sources, STAN and National Accounts databases, March 2010.

Figure 1.12. **Top 250 ICT firms' employment trends, 2000-09**

Average number of employees, index 2000 = 100*

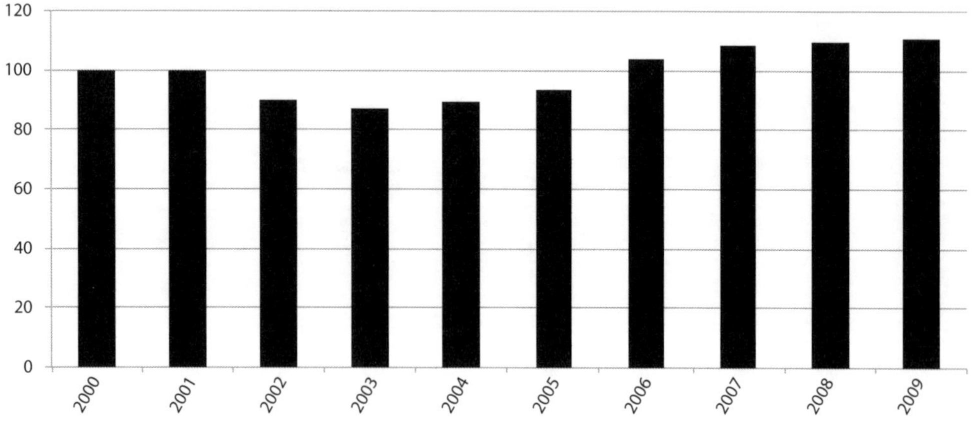

* Based on averages for those firms reporting in 2000-09.

Source: OECD, Information Technology database, compiled from annual reports, SEC filings and market financials.

point higher than total business employment growth. Finland and Sweden are the countries with the largest shares of employment in total business employment, over 8%, and these shares have increased markedly, as also happened with Luxembourg, the Czech Republic, Switzerland and Norway. The share of employment in the ICT sector declined in some countries, for example Canada and the United States.

In 2009, the top 250 ICT firms employed more than 13 million people worldwide (almost 70% of ICT sector employment in OECD countries). The average number of people employed by the top 250 firms in 2009 was almost 54 000 people. After the dot.com bust in 2001, average employment started to increase in 2004, surpassing the 2000 level only in 2006. Despite the 2009 financial and economic crisis, employment among the top 250 firms increased by 1% compared to 2008, but in many cases this was due to mergers and acquisitions, in particular in the IT equipment industry.

Between 2000 and 2009, employment among top Internet firms has grown the fastest (by 21% a year), followed by IT equipment firms (14% a year) and software firms (8% a year). In 2009, despite the downturn, IT equipment, Internet, and electronics and component firms increased

Figure 1.13. **Employment trends of top 250 ICT firms by industry***

Average number of employees, index 2000 = 100

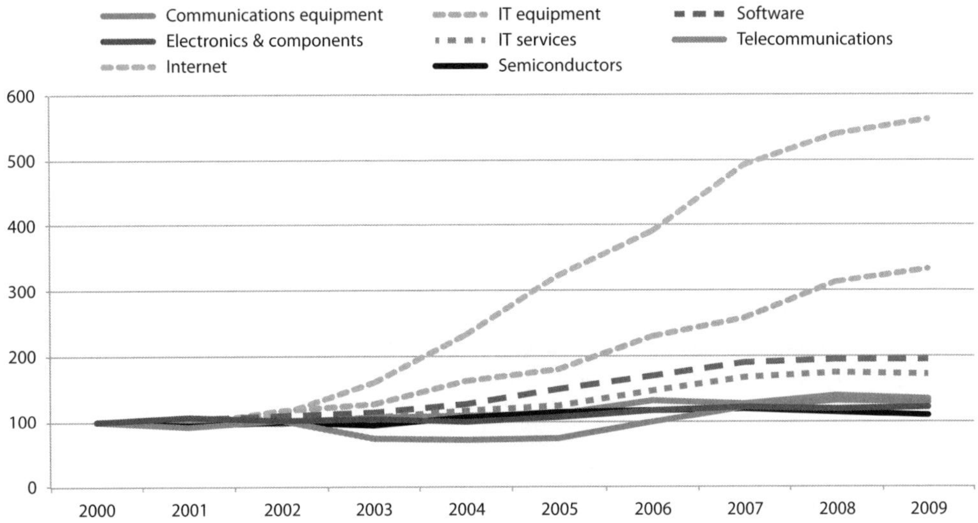

*Based on averages for those firms reporting in 2000-09.

Source: OECD, Information Technology database, compiled from annual reports, SEC filings and market financials.

employment by respectively 6%, 4%, and 2% on average. In contrast, average employment decreased most in semiconductor and telecommunication services firms by 3% and 2% respectively.

Figure 1.14. **Share of ICT-intensive occupations in the total economy, intensive users**

1995 * and 2009 **

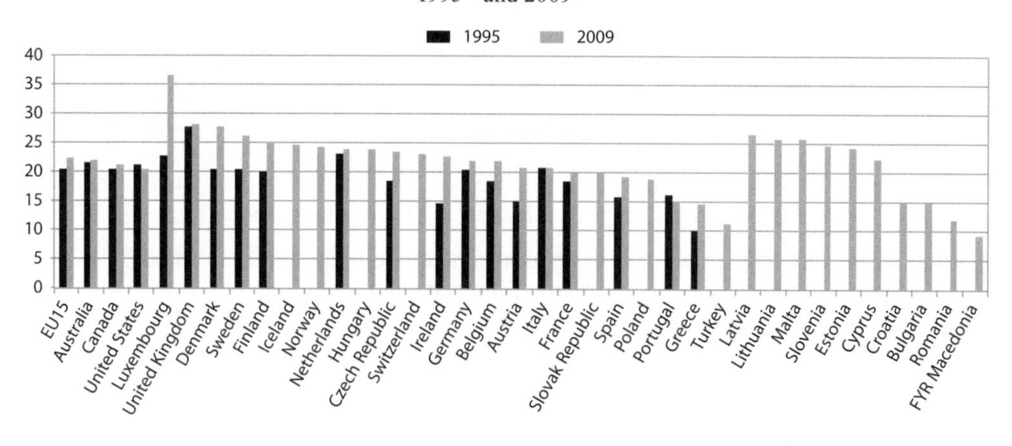

* For Australia, Finland and Sweden, 1997 instead of 1995.

** For Switzerland, the United States and Macedonia, 2008 instead of 2009. For Australia, Poland, Croatia and Malta, 2009 data are provisional as data for the fourth quarter of 2009 are not yet available.

Note: "Intensive users" corresponds to the narrow definition based on the methodology described in OECD 2004, Chapter 6. Shares for non-European countries are not directly comparable with shares for European countries, as the classifications are not harmonised.

Note by Turkey: "The information in this document with reference to 'Cyprus' relates to the southern part of the Island. There is no single authority representing both Turkish and Greek Cypriot people on the Island. Turkey recognizes the Turkish Republic of Northern Cyprus (TRNC). Until a lasting and equitable solution is found within the context of United Nations, Turkey shall preserve its position concerning the 'Cyprus issue'."

Note by all the European Union Member States of the OECD and the European Commission: "The Republic of Cyprus is recognized by all members of the United Nations with the exception of Turkey. The information in this document relates to the area under the effective control of the Government of the Republic of Cyprus."

Source: OECD calculations from EULFS, United States Current Population Survey, Statistics Canada, Australian Bureau of Statistics.

The growing importance of technology for employment in non-ICT sectors

ICT-related employment is spread widely across the economy. Many ICT employees are elsewhere in the economy carrying out ICT tasks and some employees in the ICT sector are non-ICT. ICT-using occupations are those where ICTs are used regularly as part of the job, but where the job is not focused on ICTs. ICT-using occupations make up over 20% of total employment in most countries, the exception being Eastern European countries. These occupations include, for example, scientists and engineers, as well as office workers that rely completely on ICTs to perform their tasks, but exclude for example teachers and medical specialists for whom the use of ICTs is in general not essential for their tasks. Overall, the data shows the importance of ICT – related occupations across the economy.

Around 3% to 4% of total employment in most OECD countries was accounted for by ICT specialists in 2009 with the exception being in Eastern Europe where shares are lower. This share has risen consistently in recent years in most countries, and somewhat faster than growth in the share of ICT sector employment in business sector employment (see preceding section). Among OECD ICT specialists, women still account for a relatively low share, almost 20%, with the United States, Iceland, and Finland above the OECD average.

Technology is not only increasing the demand for ICT-related professionals in ICT and non-ICT sectors, it is also changing the conditions of work in the knowledge economy. Telework is an ICT application that is changing the conditions in which a large number of professionals are providing their services. In 2006, around 23% of enterprises in the EU15 employed teleworkers, compared to only 18% in 2004. There are clear differences between northern European countries – Denmark, Norway, Iceland and Sweden – which have the highest shares of companies offering telework, and southern and eastern European countries – Italy, Poland, Spain, Hungary and Portugal – which are below the average.

Technology also has a very important role in the emergence of new business opportunities. For example, it is worth considering new forms of advertising, or the commercialisation of communication services. In the past five years, advertising expenditures on the Internet have been growing faster than on any other medium. Already in 2007, Internet advertising accounted for 7% of global advertising expenditure. The Internet now attracts more than 10% of advertising expenditures in Norway, Sweden and the United Kingdom. On the other hand, both the expansion of the mobile sector and of broadband Internet have contributed to make household expenditure on communications the fastest growing item since 1995.

Figure 1.15. **Share of ICT-specialists in the total economy, specialist users, 1995[1] and 2010[2]**

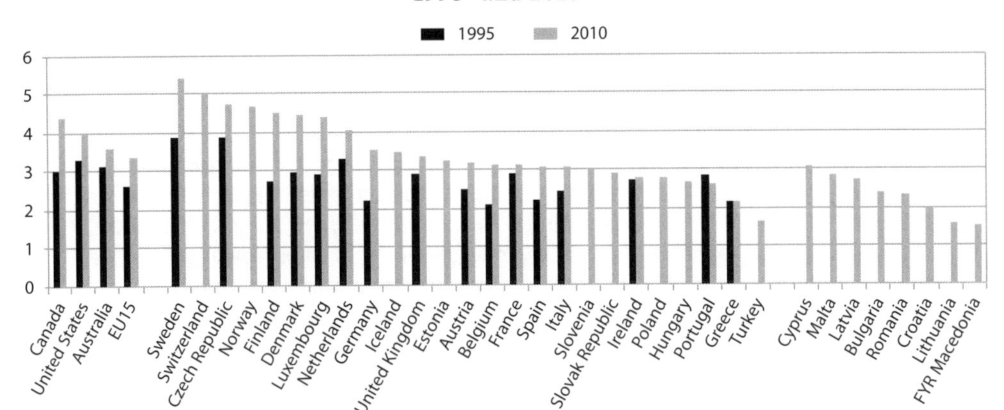

1. For Australia, Finland and Sweden, 1997 instead of 1995.

2. For Australia, 2009 instead of 2010.

Note: "Specialist users" corresponds to the narrow definition based on the methodology described in OECD 2004, Chapter 6. Shares for non-European countries are not directly comparable with shares for European countries, as the classifications are not harmonised.

Sources: Information Technology Outlook 2010 and forthcoming OECD (2011): *ICT-related skills and employment: New competences and jobs for a smarter and greener economy*, DSTI/ICCP/IE(2011)3.

Note by Turkey: "The information in this document with reference to 'Cyprus' relates to the southern part of the Island. There is no single authority representing both Turkish and Greek Cypriot people on the Island. Turkey recognizes the Turkish Republic of Northern Cyprus (TRNC). Until a lasting and equitable solution is found within the context of United Nations, Turkey shall preserve its position concerning the 'Cyprus issue'."

Note by all the European Union Member States of the OECD and the European Commission: "The Republic of Cyprus is recognized by all members of the United Nations with the exception of Turkey. The information in this document relates to the area under the effective control of the Government of the Republic of Cyprus."

Source: OECD calculations from EULFS, United States Current Population Survey, Statistics Canada, Australian Bureau of Statistics.

Figure 1.16. **Share of enterprises employing teleworkers, EU15**

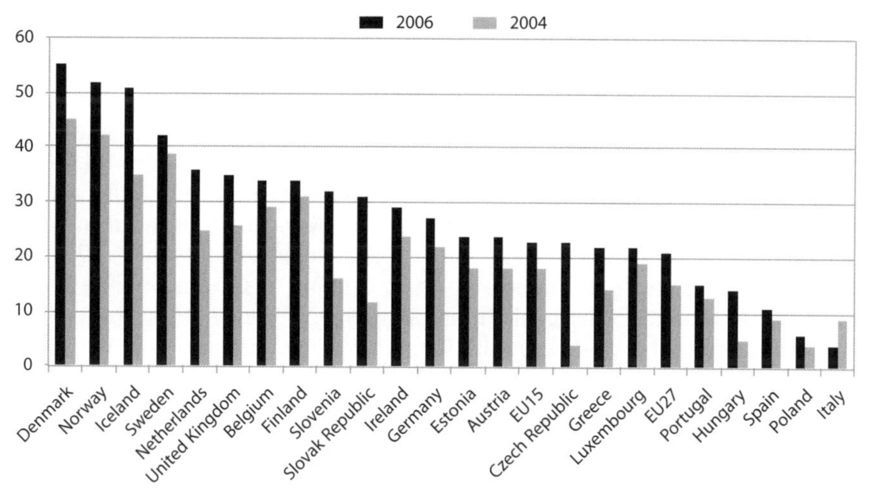

Notes: Telework is defined to include any remote location. However, the majority of teleworkers access company IT systems from home.

EU15 comprised the following 15 countries: Austria, Belgium, Denmark, Finland, France, Germany, Greece, Ireland, Italy, Luxembourg, Netherlands, Portugal, Spain, Sweden, and the United Kingdom.

Source: Eurostat survey on computers and the Internet in households and enterprises.

The need for new skills in the knowledge economy

The emerging knowledge economy, so permeated by technology, not only transforms business but also shapes the labour market by having an impact on skills demands. The transition from a manufacturing economy to a knowledge economy requires that the workforce reaches higher levels of education and develops new sets of skills compared to those required in the industrial era.

These new skills are often referred to as 21st century skills. It would be misleading to equate them with digital or technology-related skills as they go far beyond the digital domain by emphasizing new profiles of workers and new tasks. As already stated in an American report, the world resulting from the emergence of a knowledge economy "… is a world in which comfort with ideas and abstractions is the passport to a good job, in which creativity and innovation are the key to the good life, in which high levels of education – a very different kind of education than most of us have had – are going to be the only security there is." (National Center on Education and the Economy, 2006).

As already discussed in a recent OECD working paper (OECD 2009), the definition of these new skills still remains an open issue. On the one hand, there are a number of critiques which challenge the basic assumption underlying the concept of 21st century skills: namely, that the mastery of discipline-related contents is far less important for the knowledge economy than the development of complex non-routine tasks. As said in an independent American report, "it is an emphasis on what students can do with knowledge, rather than what units of knowledge they have, that best describes the essence of 21st century skills". On the other hand, although the promising concept of 21st century skills seems to be already well known and established in the public discourse, many would agree that its actual meaning remains ill-defined. Despite many inspiring attempts (Trilling and Fadel, 2009; DeRuvo and Silvia, 2010), teachers and schools may still find it difficult to see the real implications of this public discourse on their daily professional practice. This is why it may actually be crucial to start by closely examining not only what skills should be developed but also how to assess them.

As the well known analysis by Levy and Murname pointed out, basic skills, once in high demand for workers, are no longer what matter most. As reflected in Figure 1.17, there are fewer tasks requiring only routine skills and they are often done by computers. The demand for routine manual, non-routine manual and even routine cognitive tasks has decreased, while the demand for expert thinking and complex communication has increased significantly over time. Nearly every segment of the workforce now requires employees to know how to do more than simple procedures – they look for

Figure 1.17. **Evolution of occupations by the complexity of tasks in the United States**

1969-99

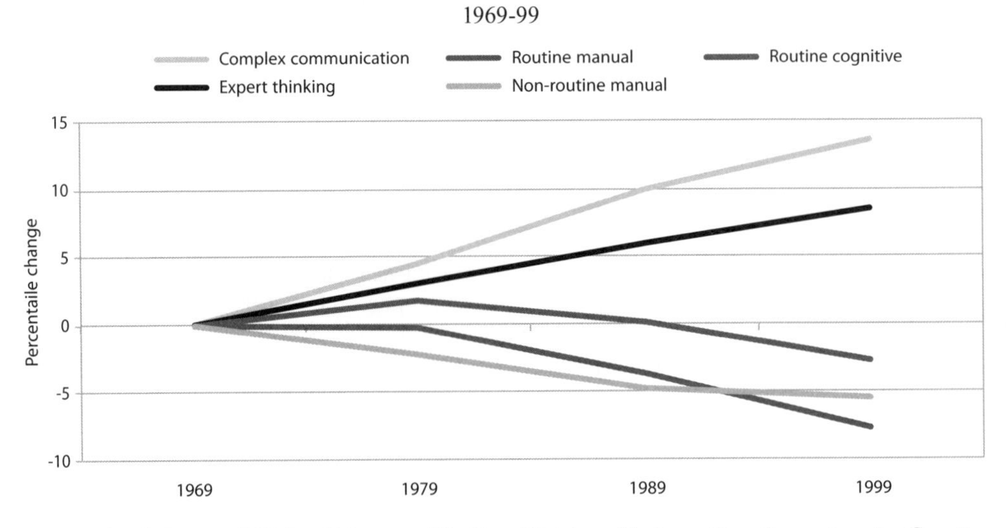

Source: Frank Levy and Richard Murnane, *The New Division of Labour: How Computers Are Creating the Next Job Market*, Princeton University Press, 2004.

workers who can recognise what kind of information matters, why it matters, and how it connects and applies to other information.

There are data to suggest that employers are increasingly interested in the so-called 21st century skills, such as communication skills, critical thinking, creativity and innovation. For example, a recent survey of more than 2000 American managers conducted by the American Management Association found that these skills and competencies are considered priorities for employee development, talent management and succession planning. In fact, the majority of the respondents agreed that their employees are measured in communication skills (80.4%), critical thinking (72.4%), collaboration (71.2%), and creativity (57.3%) during annual performance appraisal and that job applicants are assessed in these areas during the hiring process. In addition, the majority of respondents (75%) thought that these skills will become more important within their organisation in the next three to five years (American Management Association, 2010).

However, the validity of these data may be limited due to the fact the respondents to this survey were not presented with alternatives; for example, they were not asked whether they found these skills more important than others, such as technical or job-related ones.

A recent, much larger survey from the United Kingdom (England) shows that at least in this country gaps[1] and shortages are reported more in connection to job-specific skills, rather than skills such as problem solving or teamwork (UK Commision for Employment and Skills, 2010). When describing the skills lacking among their staff, employers generally focused on technical, practical or job-specific skills: almost two-thirds (64%) of employees described by their employers as lacking full proficiency were felt to lack these skills. Further, there has been an increase in the amount of skills gaps concentrated in technical, practical or job-specific skills areas over the last few years, up from 51% in 2007, 44% in 2005 and 43% in 2003.

Employers were also likely to report gaps in skills similar to the ones defined as 21st century skills, such as customer-handling and team working (reported as lacking in 50% of employees who are not fully proficient, while other skills, such as communication and problem solving ones were the next most commonly identified (46% of employees not fully proficient).

When it comes to skills shortage vacancies,[2] once more technical, practical and job-specific skills continue to be lacking in a large number of cases (62% of skill-shortage vacancies, up from 52% in 2007). Several "softer" skills are the next most likely to be lacking when recruiting, including customer-handling skills (41% compared to 32% in 2007), problem-solving skills (38% compared to 29% in 2007) and team working skills (37% compared to 26% in 2007).

The social uses of technology

Seen from the perspective of the individual, technology has brought important changes in the way people manage information, access services and communicate. According to the findings of a recent Korean survey (National Internet Development Agency, 2008), what most people (80%) found as the most important benefit of using the Internet in daily life is the overall sensation that it results in convenience – in other words, that it facilitates the management of many important aspects of living.

The fact is that despite the huge growth in new applications, such as Internet video and social networking, an international survey in 18 countries carried out by Gartner, 2008) found that the main reason for accessing the Internet is to use e-mail and gather information, which in many respects can explain why the most important positive aspect of using the Internet is convenience in life. The same survey found that the third Internet interest worldwide was online banking. Sharing photos, videos and files came in fourth, with all respondents worldwide ranking geographic navigation services – for example Google Earth – and shopping online as fifth and sixth in importance, respectively.

Communication, the most widespread use

Along with access to information, people are increasingly using the Internet for communicating. For instance, on average 57% of adult users in OECD countries used the Internet to send e-mail or telephone in 2007. In Iceland, the Netherlands, Norway and Korea the figure was above 75%.

A great majority of citizens in OECD countries have a private e-mail address, and most of them more than one – at least one private and another attached to the workplace.

Telephoning through the Internet is also becoming increasingly popular, associated with the uptake of broadband. As an indication, between 2004 and 2007 the number of registered Skype users increased 50 times up to a total community of some 276 million users worldwide.

Access to information

The wealth of information that is at people's fingertips has grown exponentially in the last decade. In the period 1999-2008 the number of Internet hosts multiplied 13 times and there are 542 million Internet hosts nowadays.

It is interesting to look closely at what kind of information people are searching and how relevant it is for critical decision-making. In the United

States 45% of Internet users in 2005 said that the information they found helped them make major decisions in the previous two years. Interestingly, health and education were the two domains in which a majority of users were seeking information. More specifically, these decisions were related with helping someone else to cope with a major illness, find the appropriate career training and choosing a school, as Figure 1.18 shows. Roughly 7% of the users in the United States used the Internet to find job prospects, but the equivalent percentage in Finland, Denmark, Norway, New Zealand and Switzerland was higher than 20%. In the United States, the number of people relying on the Internet for major decisions increased by one third from 2003 to 2005.

Figure 1.18. **American people who said that the Internet was crucial or important at least in one of these decisions**

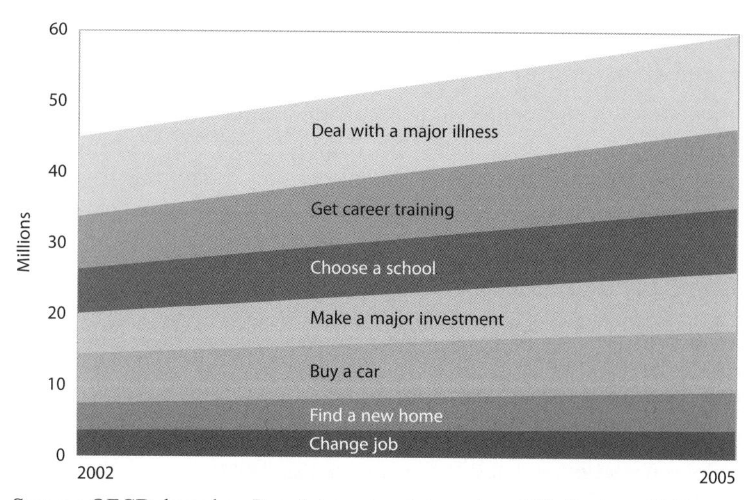

Source: OECD, based on Pew Internet and American Life Project, 2006.

The creation and sharing of content

New applications, coupled with broadband, have empowered Internet users to become content producers. Blogs, wikis, sharing sites and P2P applications make it extremely easy to upload content nowadays, such as text or various forms of media, or to contribute to existing content. For instance, 18% of all Internet users in OECD countries have created web pages in 2007 – in Korea and Iceland one user out of three did so (OECD, 2007).

Similarly, when it comes to peer-to-peer file sharing on average 17% of Europeans (EU15) use these services, with Spain doubling the European average. One adult out of four downloads music or plays games on the Internet. In Korea and the Netherlands it is almost one out of two.

Access to public and private services and goods

The Internet is also having an impact on the way people access all kinds of services or goods, be that to trade, buy or sell goods or manage financial activities such as banking, or to benefit from basic services such as education or health, provided in alternative ways, or simply to participate in government and interact with public administrations.

Trading

E-commerce is growing steadily. In 2007 over 25% of adults in OECD countries used the Internet to order or buy private goods or services. In Japan, one out of two adults ordered or purchased something through the Internet. In Norway, the United Kingdom, Denmark, Germany, the Netherlands, Australia and Korea the percentage of adults using the Internet to buy was higher than 40%.

E-banking is yet another example of how the Internet is changing the traditional behaviour of consumers. On average, over 30% of people in OECD countries use banking services on the Internet. E-banking is particularly mainstream in the Nordic countries, where over 50% of adults make use of it.

Public goods and services

Education and health are probably the two best examples of how technology changes the way in which citizens access public goods, and also how their provision can be improved.

As mentioned above, seeking critical information on health is becoming one of the most frequent uses of the Internet. In 2007, 45% or more of adults in Luxembourg, Finland, the Netherlands, Iceland and Germany sought health information on the Internet, as did over 35% in Denmark, Canada and Norway. But also from the perspective of the supply, the range of possibilities for digital delivery of health care related services is rapidly expanding, ranging from communication of patient files to distance consultation or even small surgery.

Education is another good example of how technology can improve delivery and boost quality. The area where technology has brought a more radical change is distance education ore-learning. At the individual level, as early as in 2006, over 20% of Internet users declared to have benefited from some sort of formal educational activity through the Internet in the United Kingdom, Turkey, Greece, Hungary and the Netherlands, as shown in Figure 1.19; in Finland the corresponding percentage was higher than 30%. At firm level, over 25% of enterprises in the OECD used e-learning applications for employee training. But also formal education institutions are getting better access to technology, as governments are investing increasingly in the infrastructure required. As shown in a previous report (NML PISA, 2010) more than 95% of primary and secondary schools had a broadband connection in the United States, Korea, the Netherlands and Denmark – although actual use does not seem to match the initial expectations.

Figure 1.19. **Businesses using e-learning applications for training and education of employees (2007) and Internet users declaring to use it for some form of formal education activity (2006)**

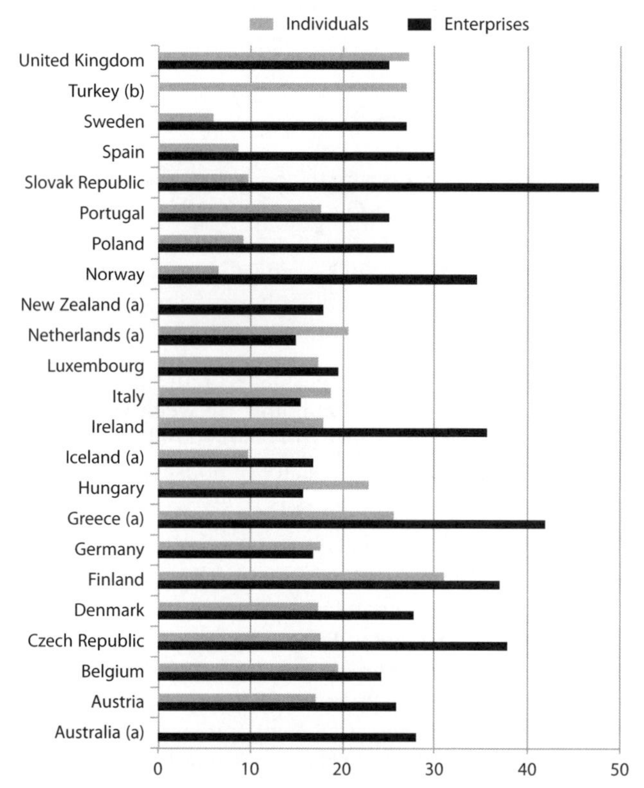

Notes:

Enterprises using e-learning applications for training and education of employees, in 2007, as a percentage of businesses with 10 or more employees.

Individuals who used Internet, in the last 3 months, for formalised educational activities (school, university, etc.) in 2006, as a percentage of Internet users.

Source: OECD, ICT database and Eurostat, Community Survey on ICT usage in households and by individuals, January 2008.

E-government

Technology seems to be improving the relationships between citizens and governments, judging by the number of convenient services and applications that are being offered under the umbrella of e-government. Moreover, technology is also expected to increase political engagement and participation.

In 2007 on average almost one-third of all adult citizens in OECD countries used the Internet for interacting with public authorities (OECD, 2008). The type of interactions range from simply obtaining information from official websites to filling in forms online. One area in which the increase of use has been spectacular in the past few years is online tax declarations.

Beyond access: new digital divides

Two main factors can explain why the opportunities offered by technology appear to become so widespread. On the one hand, the fact that technology devices, and particularly access to the Internet, has become part of the home landscape of appliances – a computer connected to the Internet seems to be seen increasingly as a must, as a mobile phone is for the individual. On the other, broadband access is in fact boosting use.

As Figure 1.20 shows, on average, 58% of the households in OECD countries have access to the Internet at home. Between years 2000 and 2007 growth in access was spectacular. By 2007, Korea (94%), Iceland (84%) and the Netherlands (83%) had the highest share of households with access to the Internet. Overall, universal home access is likely to become reality in some OECD countries in less than five years.

The development of broadband drives somehow the latest developments, allowing users to access high demanding applications or services. As a matter of fact, broadband prices have been decreasing over the past five years while speeds have stopped increasing. For instance, over 2006/2007 the speed of DSL and cable broadband increased by 29% and 27% respectively, while at the same time prices went down by 19% and 16%. Not surprisingly, broadband is quickly replacing dial-up Internet: in the OECD area the number of Internet broadband subscribers increased eleven times between 2000 and 2006 and it is the preferred way of access by the majority of subscribers (64% in the OECD area).

Opportunities are not equally distributed

It would be misleading to conclude that all these opportunities brought about by technology are equally distributed in society. Clearly, OECD countries benefit more from these opportunities than most developing countries. But even inside OECD countries not all citizens are benefitting equally from

Figure 1.20. **Households with access to the Internet**

Notes:

Generally, data from the EU Community Survey on household use of ICT, which covers EU countries plus Iceland, Norway and Turkey, relate to the first quarter of the reference year. For the Czech Republic, data relate to the fourth quarter of the reference year.

Internet access is via any device (desktop computer, portable computer, TV, mobile phone etc.).

For Australia: data was based on a multi-staged area sample of private and non-private dwellings, and covers the civilian population only. Households in remote and sparsely settled parts of Australia are excluded from the survey.

For Australia: data for 2001 and for 2004 onwards is based on a financial year, data provided relate to the second half of the reference year and the first half of the following year.

For Canada: Statistics for 2001 and every other year thereafter include the territories (Northwest Territories, Yukon Territory and Nunavut). For the even years, statistics include the ten provinces only.

For Israel: The Internet question does not distinguish between broadband and other bandwidth Internet channels.

For Japan: Internet access through mobile phone, TV and game consoles are also included.

For Korea: As of 2004, Internet access through mobile phone, TV and game consoles are also included.

For New Zealand: The information is based on households in private occupied dwellings. Visitor-only dwellings, such as hotels, are excluded.

Source: OECD, ICT database and Eurostat, Community Survey on ICT usage in households and by individuals, November 2011.

technology. There are important concerns related to a number of variables, the three most important being socio-economic status, gender and age.

Socio-economic inequalities

To begin with, education matters a lot. Technology usage is lower for less educated individuals or, to say it differently, the more educated the more intense the use of technology. The most recent data from the United States, as reflected in Figure 1.21, show clearly how this relationship operates and to what extent education and Internet usage are closely correlated: while 94% of college graduates use the Internet, this is the case in only one third of those individuals not having graduated from high school – that is, individuals with low education levels are three times less likely to use the Internet than higher education graduates.

Figure 1.21. **Individuals using the Internet in the United States by education level**

2008

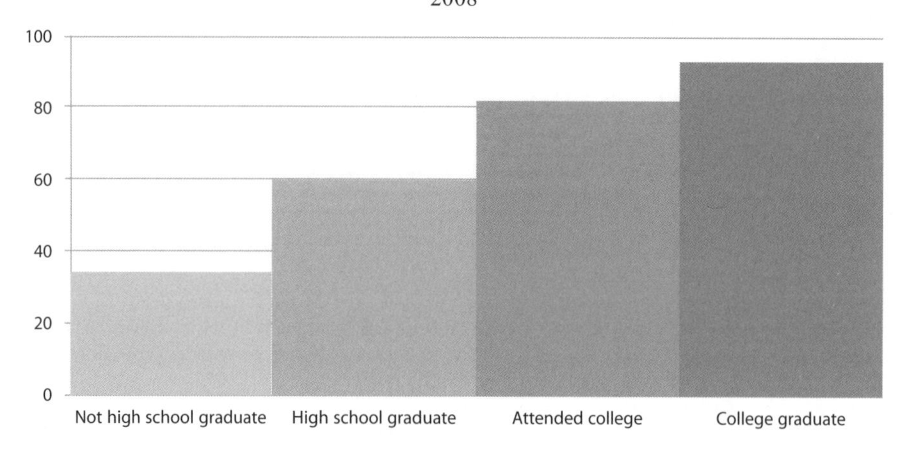

Source: Mediamark Research Inc., New York, NY, Multimedia Audiences, fall 2008.

According to the latest available data for the OECD countries (2007), the gap between individuals with higher education and those with primary education amounts to roughly 40 percentage points. The difference is particularly noteworthy in countries like Portugal, Korea, Greece, Italy, Ireland, Hungary and Spain as it is higher than 50 percentage points, as reflected in Figure 1.22. In the Nordic countries, Mexico and New Zealand such a difference is lower than 20 percentage points.

A second way to look at the impact of socio-economic status on technology usage is to highlight the differences in the percentage of users according to income. The same figure (Figure 1.22) depicts the differences in the percentage of Internet users between households with high and low incomes (fourth and first quartiles). Although data exist for only a few countries, on average there are 40 percentage points of difference between households with incomes in the fourth and first quartile. The largest differences are in Poland, the Slovak Republic and Hungary. But the most interesting aspect of this figure is that it reveals that there are countries where

Figure 1.22. **Difference between the percentages of individuals using the Internet with higher and lower levels of education and between households with high and low income**

2007

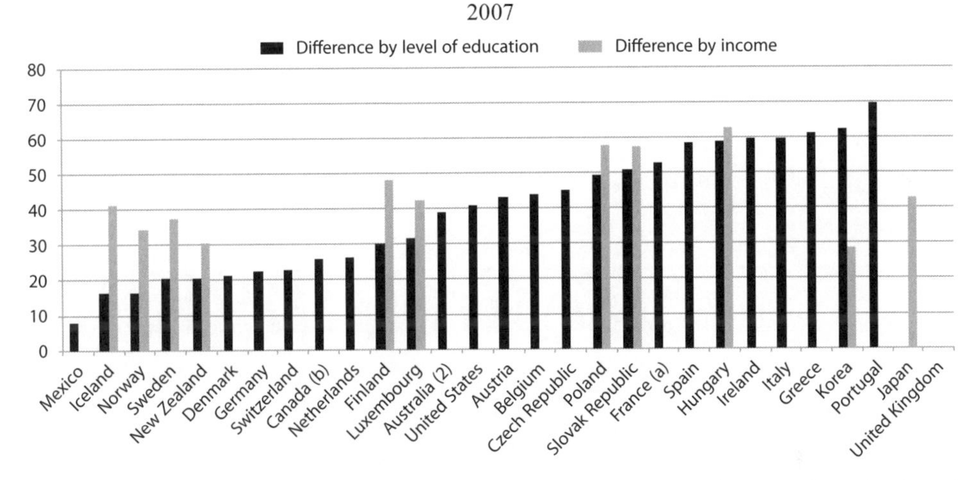

Source: OECD, ICT database and Eurostat, Community Survey on ICT usage in households and by individuals, January 2008.

Figure 1.23. **Individuals using the Internet in the United States by income level (in USD)**

2008

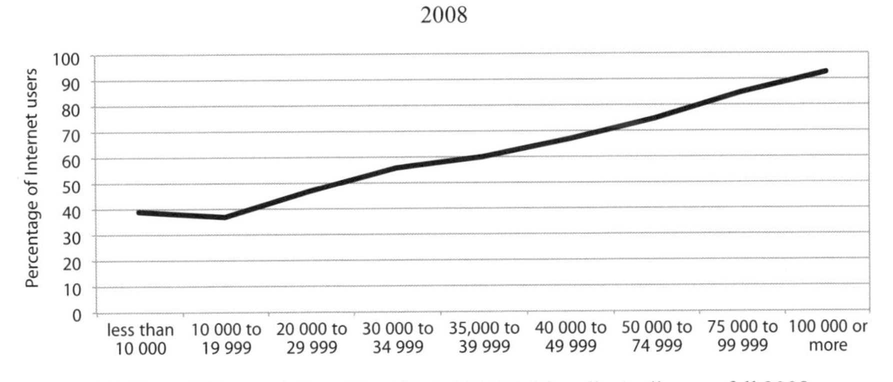

Source: Mediamark Research Inc., New York, NY, Multimedia Audiences, fall 2008.

income is a less important determinant of technology use than education (Korea), others where both income and education are equally important (Poland, Hungary, Slovak Republic) and finally countries where income matters far more than education (Iceland, Norway, Sweden, Finland, New Zealand).

The relationship between income and Internet usage can be seen with more detail in the particular case of the United States. Figure 1.23 shows how the intensity of use increases with income in a linear way: the higher the income the higher the intensity of Internet use.

Figure 1.24. **Individuals using the Internet from any location by gender, 2010 or latest available year, as a percentage of adults.**

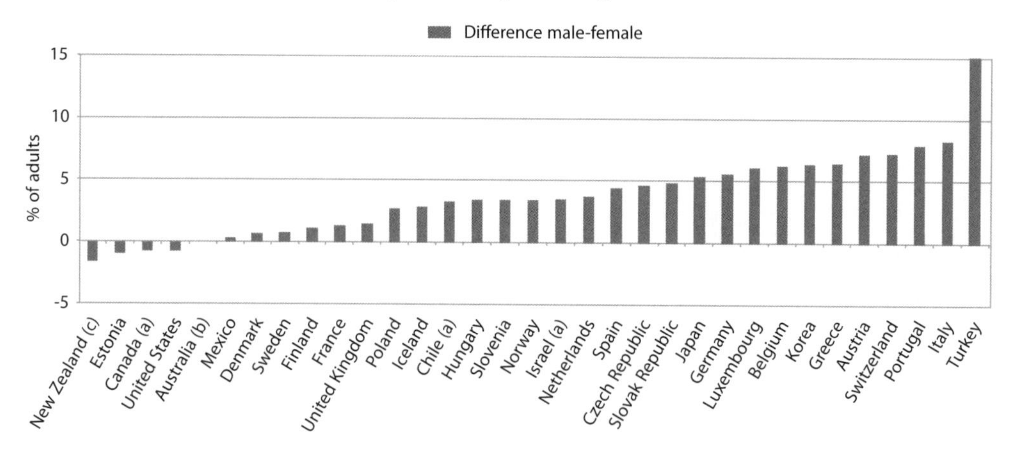

Figure notes: a. 2009; b. 2008; c. 2007; d. 2006.

Data generally refer to Internet use in the last 12 months for non-Eurostat countries and last three months for countries covered by Eurostat. Individuals aged 16-74 years, except for Israel (20-74) and Japan (6+).

Country notes:

For Israel: Data refer to the use of the Internet in the last 3 months.

For Switzerland: Data refer to Internet users who used the Internet at least once within the last six months.

Source: OECD, ICT database and Eurostat, Community Survey on ICT usage in households and by individuals, April 2011.

Gender inequalities

When it comes to gender, adult males tend to use technology more than females. However, in the United States, Canada and New Zealand the contrary is true and women use the Internet slightly more than men. As a matter of fact, the gender divide in the United States seems to be widening over time, with increasingly more women using the Internet than men – comparing data from 2008 to 2007 ((Mediamark Research and Intelligence, 2008).

Age inequalities

Finally, the age divide is quite controversial. At first glance, data shows clearly that elderly people tend to use technology less than young people. As shown in Figure 1.25, in OECD countries the average differences between young adults (aged 16 to 24) and the elderly (aged 65 to 74) are really astonishing: while an overwhelming majority of young adults are Internet users, exactly the opposite can be said about the elderly. In some countries the percentage of young adults using the Internet was higher than 95% already in 2007 – for example in all Nordic countries except Sweden, Korea, the Netherlands, Japan and Luxembourg.

Figure 1.25. **Individuals using the Internet from any location, by age group**

2007 or latest year

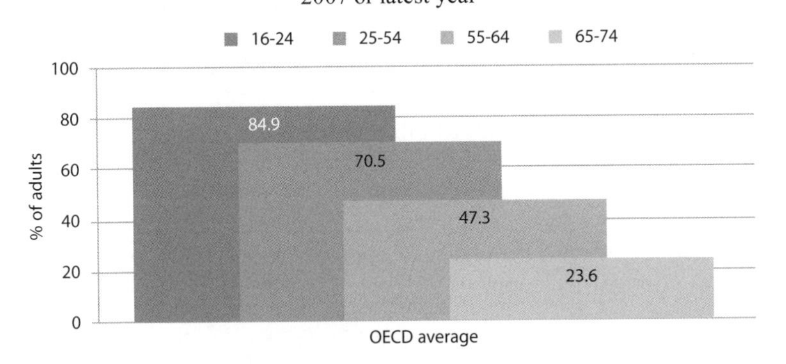

Source: OECD, ICT database and Eurostat, Community Survey on ICT usage in households and by individuals, January 2008.

Countries like the Slovak Republic, Spain, Korea and Poland show the largest divide according to age, while the difference is less acute in the Nordic countries, the United States and New Zealand. As a matter of fact, this behaviour has to be interpreted in the wider context of media consumption. Patterns of consumption are different according to age, as Figure 1.26 shows for the United States. Older individuals tend to be more attached to television and particularly to newspapers than younger adults, who combine those two with radio and the Internet thus presenting a far richer and more complex pattern of media consumption, possibly involving a high degree of multitasking. In many respects, there is a higher consumption of media until approximately the age of 45, which seems to mark a point of change. Clearly, from that point onwards individuals seem to be more attached to television and newspapers.

Figure 1.26. **Consumption of selected media in the United States, by age (2008)**

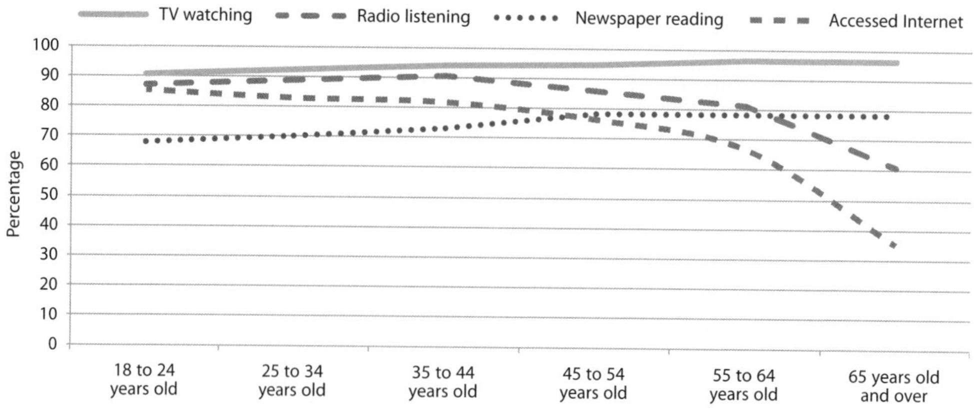

TV watching
Radio listening
Newspaper reading
Accessed Internet

Percentage
100
90
80
70
60
50
40
30
20
10
0

18 to 24 years old
25 to 34 years old
35 to 44 years old
45 to 54 years old
55 to 64 years old
65 years old and over

Source: Mediamark Research Inc., New York, NY, Multimedia Audiences, fall 2008. The percentage refers to individuals having used each media during the prior week.

Some claim that these striking differences are generational by nature – that is, that belonging to the digital native generation is what drives the attachment to technology, rather than age *per se*. Nevertheless, there are clear indications that the percentage of elderly people using the Internet is growing everywhere. At the same time, even considering that those differences originate from different generational approaches to technology, in the long run they are bound to disappear.

Notes

1. Skill gaps exist when the employer indicates that staff at the establishment are not fully proficient at their jobs. NESS records only whether staff are fully proficient or not.

2. These are defined as a subset of hard-to-fill vacancies where the reason given for the difficulty filling the position is a low number of applicants with the required skills, work experience or qualifications.

References

American Management Association (AMA), (2010), "Critical Skills Survey", *http://www.amanet.org/news/AMA-2010-criticalskills-survey.aspx*

DeRuvo and Silvia L., (2010), *The Essential Guide to RTI: An Integrated, Evidence-Based Approach*, Jossey-Bass, San Francisco, USA.

Gartner (2008), "Dataquest Insight: Consumers' Value Perception of the Internet", April 2008.

Mediamark Research and Intelligence (MRI), (2008), "Multimedia Audiences", Magazine Audience Estimates, New York.

National Center on Education and the Economy (2006), *The New Commission on the Skills of the American Workforce, Tough Choices or Tough Times,* Jossey-Bass, Washington, DC.

National Internet Development Agency (NIDA), (2008), "Survey on the Computer and Internet Usage", Korea.

OECD (2007), *Participative Web and User-Created Content: Web 2.0, Wikis and Social Networking*, OECD Publishing, Paris.

OECD (2008), *Information Technology Outlook 2008*, OECD Publishing, Paris.

OECD (2010), *Are the New Millennium Learners Making the Grade? Technology Use and Educational Performance in PISA 2006*, Centre for Educational Research and Innovation (CERI), OECD Publishing, Paris.

Trilling, B. and Fadel (2009), In Partnership for 21st Century Skills (Ed.), *21st century skills. Learning for life in our times*, John Wiley and Sons, San Francisco.

UK Commission for Employment and Skills (UKCES), (2010), *National Strategic Skills Audit for England, Skills for Jobs, Today and Tomorrow.*

Chapter 2

How relevant connectedness is for young people

Young people's attachment to digital media and connectivity will shortly reach a level of almost universal saturation in OECD countries. In the Nordic countries, the Netherlands, the United Kingdom and Austria, more than 95% of 15-year-olds use a computer connected to the Internet daily while at home. On average, two hours per day are devoted to a number of ICT activities, mostly related to social interactions and the consumption of digital content, sometimes in connection with school-related tasks. However, despite these impressive developments the use of a general stereotype, such as the New Millennium Learners, may be inappropriate when it comes to understanding the complexity of the implications that digital media and connectivity can have on the lives on young people. Only a higher level of granularity can unveil important differences among learners that often translate into alternative profiles, needs and expectations in relation to both technology and learning.

Note by Turkey: "The information in this document with reference to 'Cyprus' relates to the southern part of the Island. There is no single authority representing both Turkish and Greek Cypriot people on the Island. Turkey recognises the Turkish Republic of Northern Cyprus (TRNC). Until a lasting and equitable solution is found within the context of United Nations, Turkey shall preserve its position concerning the 'Cyprus issue'."

Note by all the European Union Member States of the OECD and the European Commission: "The Republic of Cyprus is recognised by all members of the United Nations with the exception of Turkey. The information in this document relates to the area under the effective control of the Government of the Republic of Cyprus."

The statistical data for Israel are supplied by and under the responsibility of the relevant Israeli authorities. The use of such data by the OECD is without prejudice to the status of the Golan Heights, East Jerusalem and Israeli settlements in the West Bank under the terms of international law.

Although an increasing percentage of young people can be said to be adept in technology and permanently connected, it is misleading to assume that all of them fit equally well into the usual stereotypes suggested by powerful images such as the NML or digital natives. As is the case with learning styles, there are different profiles of children and young people regarding technology adoption and use, and in many respects clear digital divides still exist. The use of concepts such as the NML can be helpful in so far as they evoke a clear and powerful image that is extremely suggestive. But when these concepts are also used as clichés or stereotypes from which lessons can be learnt, they become misleading. For the purposes of improving teaching and learning in formal education, the diversity of students' needs and expectations and situations matters most.

Most young people in OECD countries are fully connected

An increasing percentage of children born in OECD countries grow up in societies where Internet connections, mobile phones and video game consoles are readily available to them. Children are often recognised as one of the most important driving forces behind the acquisition of technology devices. This applies to a vast array of them, ranging from game consoles to computers, as well as to communication services and particularly to the Internet. For instance, there is evidence showing that households with children are more likely to have access to the Internet than those without. On average, in OECD countries 73% of households with children had access to the Internet, compared to only 53% of households without children. As shown in Figure 2.1, this is quite clear in Finland, France, Germany and Belgium. Korea, the United Kingdom and the United States had the smallest difference. In fact, it has been demonstrated that the presence of children in the home may be a primary reason for the adoption of computers and the Internet in the household.[1]

Recent OECD work on the NML has demonstrated (OECD, 2010) that in a number of OECD countries, namely the Nordic countries, the Netherlands, the United Kingdom and Austria, more than 95% of 15-year-olds use a computer connected to the Internet daily while at home. On average, two hours per day are devoted to a number of ICT activities, mostly related to social interactions and the consumption of digital content, sometimes in connection with school-related tasks.

The latest PISA data (2009) reinforce this image, particularly in three areas: access to computers, access to mobile phones, and access to the Internet.

On average, for the 25 OECD countries for which data are available the percentage of 15-year-olds never having used a computer is very low, only 0.8%. In countries like Korea or Finland the percentage is nil, that is there was not one student participating in the sample who had never used a computer. Greece, Turkey and Japan show the highest percentages, with figures above 2%, as Figure 2.2 shows.

Figure 2.1. **Household Internet access by household type**

2007

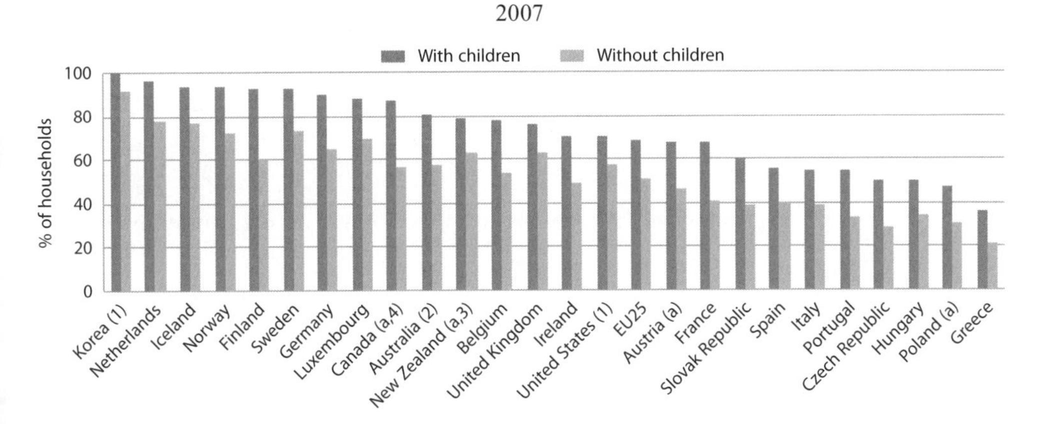

Figure notes:

a. 2006; b. 2005; c. 2004; d. 2003.

Generally, data from the EU Community Survey on household use of ICT, which covers EU countries plus Iceland, Norway and Turkey, relate to the first quarter of the reference year. For the Czech Republic, data relate to the fourth quarter of the reference year.

Country notes:

1. Households with dependent children are defined as households with children under the age of 18.
2. Data provided relate to households with or without children under 15 years.
3. The information is based on households in private occupied dwellings. Visitor-only dwellings, such as hotels, are excluded. Household child dependency status does not include households where there is a child with an unknown dependency status.
4. Dependent children refers in the survey to single, never married children of the household reference person, of any age. Statistics for 2001 and every other year thereafter include the territories (Northwest Territories, Yukon Territory and Nunavut). For the even years, statistics include the ten provinces only.

Source: OECD, ICT database and Eurostat, Community Survey on ICT usage in households and by individuals, January 2008.

The percentage of 15-year-olds in the same countries who do not have access to a mobile phone is far higher: 5%. This challenges the assumption that all young people have a mobile phone. Yet, leaving aside the cases of Turkey and Canada, which in this respect can be considered as outliers, the average percentage for the rest of countries goes down to 3%. The differences in access to computers and to mobile phones are likely to be based on the fact that mobile phones require some form of regular fee.

With the development of applications that hold enormous appeal for young people – especially social networking sites such as Facebook or video sites such as YouTube – the amount of time children and young people spend on a computer in a typical day for recreational purposes has increased exponentially. According to a recent American survey (Kaiser Family Foundation, 2010) this amount of time has doubled over the past ten years: from an average of 58 minutes per day in 1999 to 2 hours and 17 minutes in 2009.

Figure 2.2. **Percentage of 15-year-olds who have never used a computer**

2009

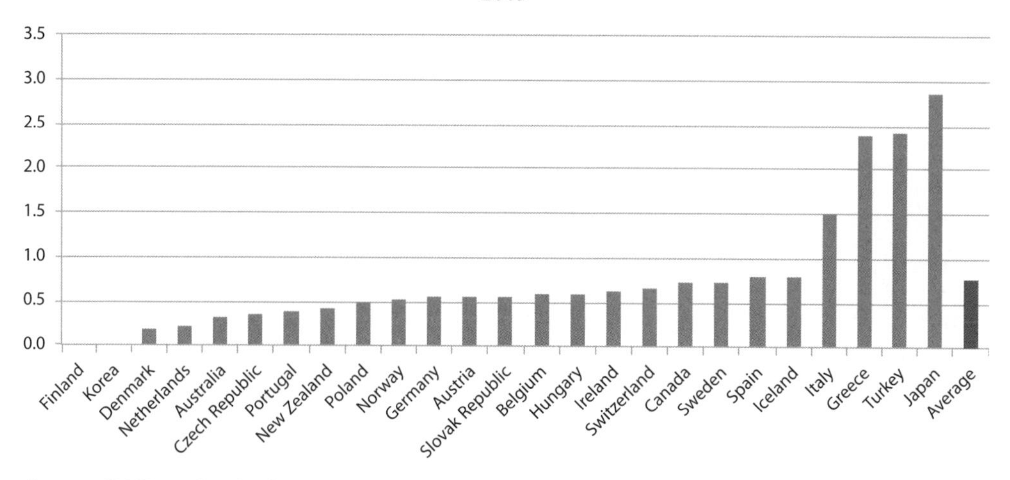

Source: OECD, PISA database, 2010.

Figure 2.3. **Percentage of 15-year-olds who do not have a cell phone**

2009

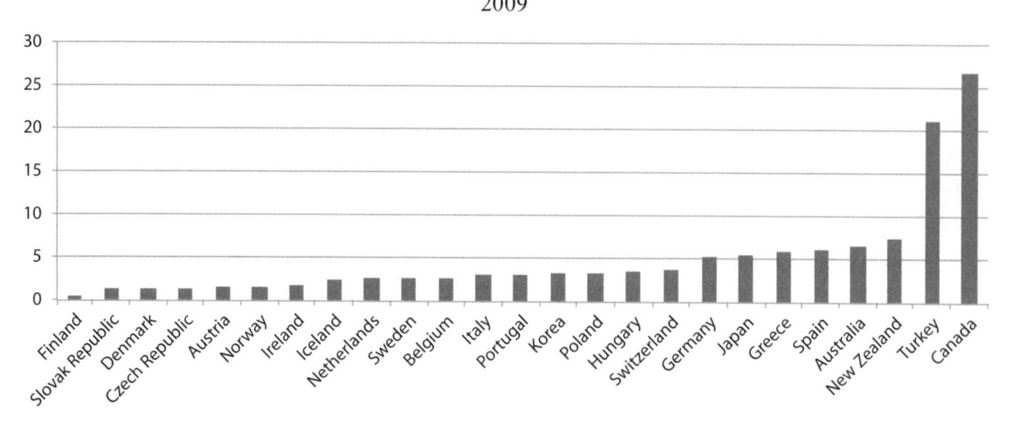

Source: OECD, PISA database, 2010.

Increasingly, it becomes clear that the main use of the computer is to get connected. Figure 2.4 shows the percentage of 15-year-olds who use the Internet at home and at school. Three facts become apparent. First, that on average for the 25 OECD countries for which data are available, 90% of 15-year-olds have access to Internet from home. Second, that in a number of OECD countries, access from home is almost universal; this is the case of the Netherlands, all Nordic countries, Switzerland, Belgium, Korea, Germany, Austria and Canada, with percentages above 95% and the Netherlands, Iceland and Finland with 99%. Third, that only in a minority of countries access to the Internet at the school roughly matches the percentage at home, with Denmark, Australia, Norway, the Slovak Republic and New Zealand showing matching rates of access and thus revealing different policy priorities and school policies in this domain. Contrarily, Italy, Belgium, Germany and, to a lesser extent, Korea show the greatest gaps.

Figure 2.4. **Percentage of 15-year-olds with access to the Internet, at home and at school**

2009

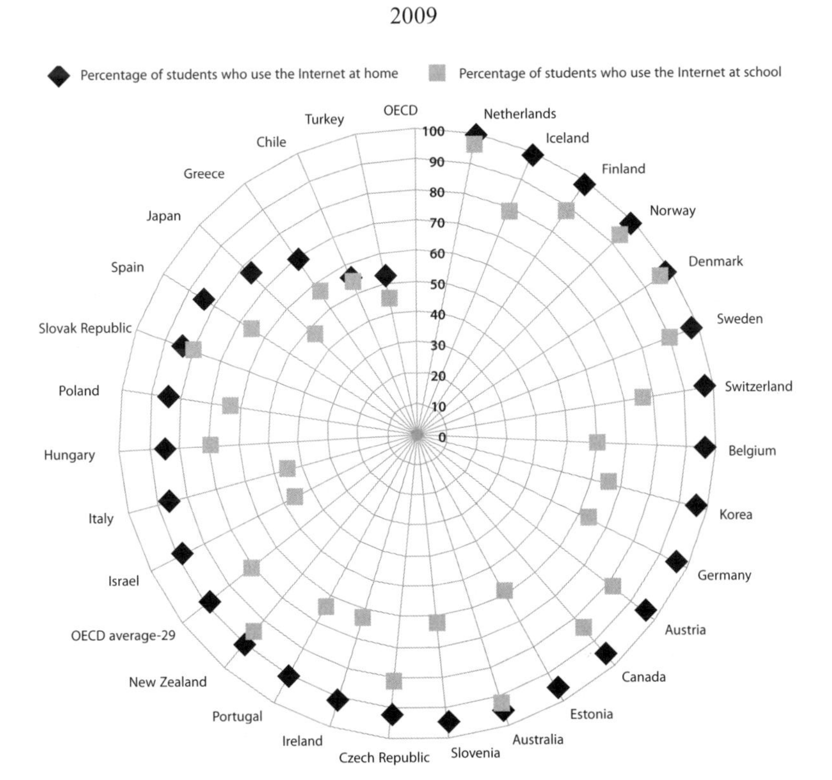

Source: OECD, PISA database, 2010.

In conclusion, it can be supported with evidence that young people are increasingly connected. In fact, there is sufficient international evidence to support the claim that the younger generations, particularly those between the ages of 15 and 24, are by far the population segment with the highest intensity of connectedness: not only the population group with the highest percentage of people online but also with the highest intensity of use – spending on average the most time connected.

Such an increase in use correlates with media preference as well. In a United Kingdom survey it was found that given a choice of six media, one-third of children aged 8 to 17 would choose the Internet as their only choice if they could not have any other, surpassing television, telephone and radio (BBC Monitoring International Reports, 2002). However, in practice, the emergence of computers and the web, partly thanks to multitasking, has not reduced exposure to other media; on the contrary, computers increase the overall "screen" time by roughly 40% (Subrahmanyam *et al.*, 2000). It is the Internet that increases the time spent with media (National Center for Education Statistics, 2004).

The computer is the epicenter of media multitasking for young people, who seem to be heavily multitasking, *i.e.* they tend to be engaged in multiple computer activities at the same time – *e.g.* texting while surfing the Internet and downloading music. In a recent American study (Kaiser Family Foundation, 2010) it was found that four teenagers in ten (40%) say they use another medium or text message most of the time, while they are using the computer; another 26% say they do so only some of the time.

The intensity and variety of uses of technology and connectedness

Going online is now thoroughly embedded in children's lives. First, an increasing number of children have now access to the Internet; they remain connected for more time, increasingly using portable devices; and more importantly, the range of activities they carry out is spreading. Children's use is increasingly individualised, privatised and mobile (Livingstone, Ólafsson and Staksrud, 2011) and with a socialisation purpose.

An increasing number of children have access to the Internet, mostly owing to the multiplication of computers in households and in schools. In the United States, 35% of public schools had access to the Internet in 1994 and 100% nine years later (Schmidt and Vandewater, 2008); home Internet access for 8-18 year-olds nearly doubled over the last ten years (from 47% in 1999 to 84% in 2009) (Kaiser Family Foundation, 2010). The percentage of children using the Internet in the European Union increased from 70% to 75% over three years (2005-08) (European Commission, 2008). 93% of American children had access to the Internet in 2007 (Pew Internet and American

Life Project, 2007). In 2006 in Japan, this was the case of 65% of children aged 10-14 and 90% of teenagers aged 15-19. In the European Union, 75% of 6-17 year-olds were reported by their parents in 2008 to use the Internet; the percentage ranged from 93-94% in Finland, Iceland and the Netherlands to 50% in Greece and 45% in Italy (Livingstone *et al.*, 2010). Ofcom's research shows that 99% of British children aged 12-15 use the Internet, 93% of 8-11 and 75% of 5-7 (Ofcom, 2010).

A second important point is that children are not only increasingly connected but also spending more time on the Internet. In 2007, British children aged 12-15 spent on average 13.8 hours a week on the Internet, nearly twice as much time as in 2005 (7.1 hours a week) (Ofcom, 2010). Already in 2003, Americans aged 13-24 already spent 16.7 hours a week on the Internet, *i.e.* more time than watching television (Kaiser Family Foundation, 2010). According to a recent European survey use is embedded in children's daily lives 60% of children go online daily or almost every day. In some countries such as Sweden, Bulgaria, Estonia, Denmark, Norway and the Netherlands, this is as high as 80%. Across Europe, 93% of 9-16 year old users go online at least weekly: children now spend on average 88 minutes per day online. 15-16 year olds spend 118 minutes online per day, twice as long as 9-10 year olds (58 minutes) (O'Neill, Livingstone and McLaughlin, 2011).

Third, what matters, however, is what children do while they are online. Their activities largely depend on age and changing usage trends. In 2007 Ofcom provided a list of 24 activities carried out by children online and classified by age range (Ofcom, 2007). Playing games was the most popular activity for children aged 8-11 but ranked fourth for children aged 16-17 after general surfing, sending and receiving e-mails and finding/downloading information for school. While 53% of 16-17 year-olds used social networking sites only 6% of 8-11 year-olds did so. Internet uses are extremely dynamic and trends in each type of use change rapidly. "Web 2.0" has modified Internet use by children, and the Pew Internet and American Life Project (Pew Internet and American Life Project, 2007) mentions that use of chat rooms decreased from 24% in 2001 to 18% in 2006. This likely reflects the fact that instant messaging functions are now an integral part of every social network or online community. In Australia in 2008, 90% of young people aged 12-17 reported using social networking services, with 51% of 8-11 year olds using these services (ACMA, 2009). According to a recent American study (Kaiser Family Foundation, 2010), visiting social networks has become the most popular activity among children aged 8-18.

Fourth, although home is still the most preferred location for Internet acces, devices are diversifying and more sophisticated mobile phones increasingly enable Internet access. The most common location of Internet use is at home (87%). For most children, this means accessing the Internet from a PC in a public room (62%), but nearly half (49%) go online in a private room where it is difficult for parents to monitor their Internet usage. Older children, boys and children whose parents have higher educational attainment are more likely to have private access from their own bedroom. Parents' Internet use also appears to be an important predictor for children's use of the Internet in the bedroom (Hasebrink *et al.*, 2011). Yet, despite this fact, 33% of European children now go online via a mobile phone or handheld device (O'Neill *et al.*, 2011), as 60% Japanese children already did back in 2007 (European Commission, 2008). It is likely that children will progressively make more use of Internet-enabled mobile devices in most OECD countries, following the Japanese example, depending on countries' socio-economic conditions: for example, 14% of British children aged 12-15 used their mobile phone to access the Internet in 2009 (Ofcom, 2010). Moreover, the age at which children acquire their first mobile phone is dropping: the Pew Research Center's Internet and American Life Project, which tracks adolescent cell phone use confirms this trend: 58% of those aged 12 owned a mobile phone in 2009 while only 18% did in 2004 (Pew Internet and American Life Project, 2009). According to another Pew Internet research, 19% of 12-17 year-olds access the Internet through portable gaming devices (Pew Internet and American Life Project, 2010). In 2009 in the United Kingdom, 12% of 5-15 years-old used their gaming console to access the Internet, rising up to 18% with children aged 12-15 (Ofcom, 2010). Children seem to access the Internet via mobile devices in addition to fixed computers rather than instead of them.

Finally, what matters is what children do online. In Europe the most popular Internet activities (Hasebrink *et al.*, 2011) are using the Internet for schoolwork and playing games alone against the computer. 14% don't get further than this, including nearly a third of 9-10 year olds and a sixth of 11-12 year olds. Next in popularity is watching video clips online (*e.g.* YouTube). These are all ways of using the Internet as a mass medium – for information and entertainment. Half of 9-10 year olds only get this far, along with a third of 11-12 year olds. In addition, most children use the Internet interactively for communication (social networking, instant messaging, email) and reading/watching the news. This captures the activities of two-thirds of 9-10 year olds but just a quarter of 15-16 year olds. More sophisticated, contact-based activities include playing with others online, downloading films and music and sharing content peer-to-peer (*e.g.* via webcam or message boards). Across Europe, over half of 9-16 year old Internet users reach this point, although only one third of 9-10 year olds and less than half of 11-12 year olds do. Finally, a quarter of children do visit chatrooms, file-sharing, blog and spend time in a virtual world. Less than one-fifth of 9-12 year olds and only a third even of 15-16 year olds do several of these activities.

Yet not all young people are equally connected

What is not very often said is that there is also strong evidence supporting the fact that not all young people are the same in this respect for two main reasons: the digital gap and digital profiles.

A digital gap still persists and a second one is emerging

There is still a digital gap even within the younger generations. To state it simply, children from households with higher educational levels have access to more locations, platforms, have more private access and more sophisticated mobile access (Helsper and Lenhart, 2011) and a significant percentage of young people are not connected and some do not even have access to a computer at home. The actual size of this excluded group varies across countries but it can be taken everywhere as a clear indication of a digital gap. This gap, the origins of which are of socio-economic nature, has not been properly tackled (OECD, 2010). Although there are clear indications that it is declining steadily in most OECD countries, the more reduced it becomes the more difficult it seems to address it.

On top of the persistence of this first digital gap, there are clear indications of a second digital divide emerging. A second form of digital divide has been identified between those who have the necessary competencies and skills to benefit from computer use and those who do not (OECD, 2010). These competencies and skills are closely linked to students' economic, cultural and social capital.

The importance of this second digital divide is clearly underestimated in education. Computer use can make a difference in educational performance if the student has the appropriate set of competencies, skills and attitudes. Teachers and schools can make a difference for students who lack the cultural and social capital that will allow them to benefit from the use of digital media in a way that is significant for their educational performance. If teachers and schools fail to acknowledge this second digital divide, and act accordingly, they will reinforce its emergence. Therefore it is important to realise that the fact that students appear to be technologically "savvy" does not mean that they have developed the skills and competencies that will make them responsible, critical and creative users of technology.

There are different profiles of learners

The second reason why it cannot be easily said that all young people are equally connected is that there are different user profiles. It is important to recognise that a proportion of young people, be they designated anti-digital or digitally-reluctant, may have opted out from being connected for personal reasons or beliefs. It should be noted, however, the elucidation of these profiles is still sociological work in progress and the number, names and distinctive characteristics of each profile are far from having gained broad consensus.

Different profiles of students *vis à vis* technology coexist as a number of studies, mostly in higher education, have demonstrated. A study at the University of Melbourne (Kennedy, G. *et al.*, 2006) noted that there is little empirical support for the stereotypical depiction of the digital native – wired and wireless 24/7. When one moves beyond entrenched technologies and tools (*e.g.* computers, mobile phones, e-mail) the patterns of access and use of a range of other technologies show considerable variation. Similarly, in a much broader study conducted in Australia four distinct types of learners were identified according to their patterns and intensity of use of technology: power users (14% of sample), ordinary users (27%), irregular users (14%) and basic users (45%) (Kennedy, G. *et al.*, 2010). Another important exception to this overall emphasis on the homogeneity of students is the Numediabios study (Ferri *et al.*, 2008), which concludes that there is enough evidence to support the existence of at least three different higher education student profiles. These profiles result from a crossed analysis combining two factors: intensity of Internet use and content production, defined as uploading content to sites like My Space, Wikipedia, YouTube and, more generally, activity in social networks, as the Figure 2.5 reflects.

Figure 2.5. **Different profiles of students according to their attachment to the Internet**

Italy, 2008

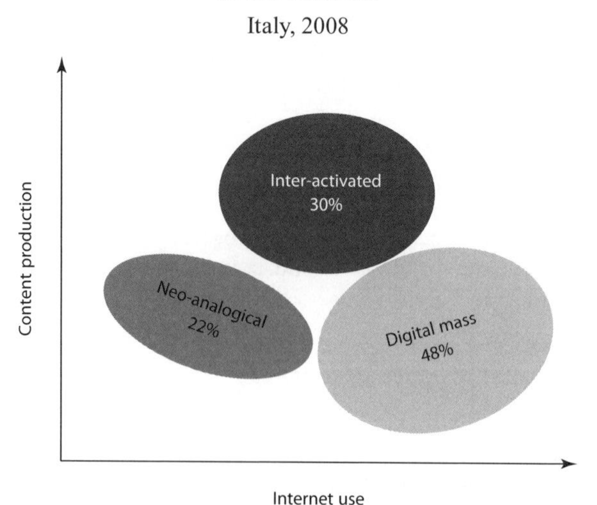

Source: Ferri *et al.*, 2008.

Such a diversity of profiles may have quite different implications at both the policy and the institutional level. The three profiles are characterised as follows:

- The **digital mass**, which accounts for almost half of the students, are heavy Internet users but not so keen on producing digital content.
- The **neo-analogical**, roughly a fifth of the students, produce some content but connect less to the Internet than the average student; in a way, they are not as dependent on Internet use as the digital mass.
- The **inter-activated**, roughly a third of the students, are the ones who better fit into the prevalent image of New Millennium Learners: heavy Internet users and quite frequent content producers.

Young people's uses of the Internet change with age

It is reasonable to assume that young people use the Internet in different ways as they grow older. In other words, it is perhaps misleading to treat all young people as one homogenous group, as their uses of technology are likely to change according to the particular developmental stage they are in. Despite this, there is little research focusing on the differences in Internet or technology use among different age groups, with most studies focusing on a particular age or treating children, teenagers and young people as one group.

To begin with, Internet use increases with age. In 2008 in the European Union, the Internet was used by 50% of 6-7 year-olds and 86% of 15-17 year-olds (European Commission, 2008). In Australia, a recent study showed that children aged 8 to 11 used the Internet on average 4.1 days per week for 1.3 hours per day, and that 12 to 17 year-olds used the Internet on average 6.3 days for an average of 2.9 hours per day (ACMA, 2009). Yet, it is true that children start to use the Internet younger. A 2009 Swedish report points out that the age of Swedish children's first use of the Internet dropped from 13 years in 2000 to 4 years in 2009. The report considers that at least half of 4-year-olds use the Internet at least occasionally (Beantin Webbkommunikation, 2010). In 2009, 74% of British children aged 5-7 had access to Internet (Ofcom, 2010). Interestingly, children are going online at ever younger ages and the average age of first Internet use is dropping across Europe. On average, children were 9 years of age when they first went online. This varies by age however: while 15-16 year olds say they were 11 on first use, younger users now say they were 7 when they started going online. There is variation across Europe: average age of first use is 7 in Denmark and Sweden and 10 in countries such as Greece, Italy, Turkey, Cyprus,[2] Demark, Austria and Portugal (Livingstone *et al.*, 2011).

One recent study that provides more detailed data in that respect is the EU Kids Online survey, funded by the European Commission's Safer Internet Programme. The survey collected data from nearly 25 000 children aged 9-16 and their parents in 25 European countries regarding Internet use (Livingstone *et al.*, 2010). As the authors point out, age differences in online activities are greater than gender or socio-economic ones. When examining the data in terms of two age groups, 9-12 and 13-16 year-olds, it is clear that they differ not just in terms of the amount of time they spent online, with the older age group using the Internet more, but also in terms of their breadth of use, with younger children doing on average five activities and teenagers eight or nine.

There is a gender gap in terms of time spent using computers, although possibly lower than it is believed. Among 8- to 18-year-old Americans, boys spend an average of about 15 minutes more per day with computers than girls (Kaiser Family Foundation, 2010). This difference is primarily due to boys spending more time playing computer games (25 minutes compared to only 8 for girls), and watching videos on sites such as YouTube (17 vs. 12). The one computer activity to which girls devote more time than boys is visiting social networking sites (25 minutes for girls vs. 19 for boys). On average, boys and girls are equally likely to visit social networking sites in a typical day (40% of each), but of those who visit, girls remain there longer (1:01 vs. :47). Interestingly, the gender difference in computer time only begins to appear in the teenage years. In other words, boys and girls start out spending equal amounts of time on a computer, but a disparity develops over time. Among 15- to 18-year-olds, there is a 42 minute gender gap (1:59 for boys, and 1:17 for girls). One clear reason for the disparity in this age group is that girls lose interest in computer games as they enter their teenage years, while boys don't. Girls go from an average of 12 minutes a day playing computer games when they are in the 8- to 10-year-old group, down to just three minutes a day by the time they are 15 to 18 years old; there is no such decrease among boys.

More specifically, 9-12 year-olds are much less likely to use the Internet for watching or posting video clips or messages, for reading or watching the news, for communication through instant messaging, email etc. and for downloading music or films compared to 13-16 year-olds. When it comes to the use of social networking sites, 59% of 9-16-year-olds have a social networking profile. This includes 26% of 9-10-year-olds, 49% of 11-12 year olds, 73% of those aged 13-14 and 82% of 15-16-year-olds (Livingstone *et al.*, 2010). Figure 2.6 clearly demonstrates that the use of social networking sites increases with age, thus following the increasing socialisation needs that grow with adolescence.

Figure 2.6. **Patterns of SNS use by age in selected countries**

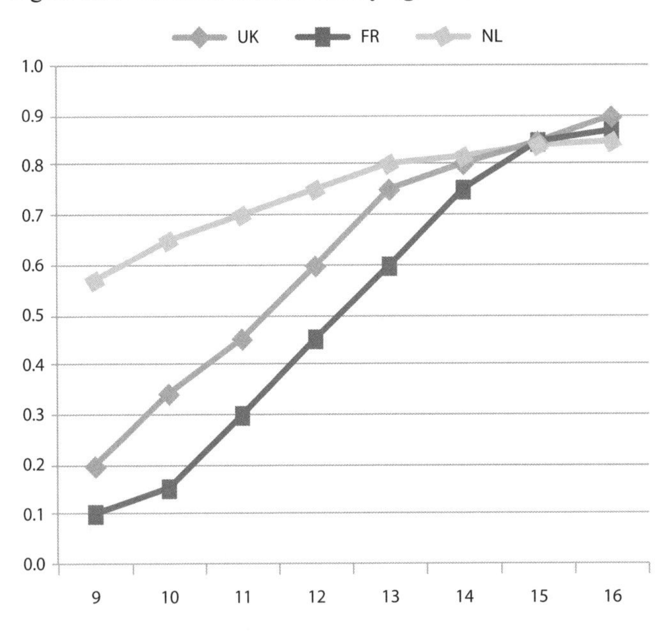

Source: EU Kids Online, 2011.

Recent data from the United States point in the same direction. A recent large-scale survey by the Kaiser Family Foundation about young people's media use examined, among other things, differences in the types of use by age groups (Rideout, Foehr and Roberts, 2010). Considerable differences are reported in the amounts of time children and young people spend using a computer for social networking and instant messaging. In the case of the former, this ranges from 5 minutes on average per day for 8-10 year-olds to 29 and 26 minutes per day respectively for the 11-14 and 15-18 age groups. For instant messaging, the equivalent amounts of time spent daily on average are 3 minutes for the youngest age group and 14 minutes for the two older age groups.[3]

Further evidence comes from an earlier survey of more than 5 000 Canadian 9-17 year-olds (Media Awareness Network, 2005). Young people were asked what they actually did online "on an average school day" and they had to choose among a range of activities, such as playing computer games, instant messaging, doing homework, etc. The results show that all activities increase with age, for both boys and girls, except for playing computer games, which declines rapidly for girls and slowly for boys. More specifically, the proportion of children who engage in instant messaging, listen to/download music, and do homework increases rapidly from Grade 4 (9-10 years), then reaches a plateau in Grades 6 (11-12 years) to 8 (13-14 years) at which point it levels off or climbs very slowly. For example, the amount of time spent talking to friends on instant messaging increases steadily from an average of 26 minutes per day in Grade 4 (9-10 years) to 68 minutes per day in Grade 11 (16-17 years).

The above data suggest that technology use changes to fit young people's developmental needs, with applications that enable them to communicate with and stay connected to their peers becoming much more widespread once they enter adolescence. This is in accordance with the well-established findings of the importance of peers to teenagers. It appears that what new technologies are providing young people is the opportunity to stay connected and socialise with their peers at the time when this socialisation becomes important to them, *i.e.* during adolescence. If this is the case, it would be interesting to see whether time spent using applications such as instant messaging, chats and social networks decreases as adolescents move into early adulthood, suggesting perhaps that high levels of use are not a cohort or generation but rather an age effect.

To sum up, there are a number of indications suggesting that today's students are most likely to be New Millennium Learners, growing steadily and already having a universal character in some OECD countries. However, there are limits to this characterisation. As mentioned above, in the same way that students have different learning styles, they also have different profiles in relation to their uses of digital media. These profiles can be drawn both in terms of access and use (Kennedy *et al.*, 2008; Oliver and Goerke, 2007; Thinyane, 2010). Moreover, students' proficiency with technology (also referred to as technological efficacy, as opposed to technological anxiety) stems from individual differences based on age, experience, exposure, personality, culture and environment (Austin, Lawson and Holder, 2007; Johnson and Howell, 2005; Joiner *et al.*, 2007). Moreover, young people make an appropriation of technology and use connectedness to suit their needs, and not the other way around. When it comes to social activities within the peer group of reference, what connectedness does is to increase the intensity of opportunities for interaction, and certainly extend them over the boundaries imposed by face-to-face exchanges. Therefore, it would be a vast oversimplification to claim that all students are represented equally by the generally accepted conception of the New Millennium Learners.

Notes

1. As Drotner pointed out as early as 2000, access to digital technologies is greater in homes with children than in those without (Drotner, 2000).

2. The following note is included at the request of Turkey: "The information in this document with reference to 'Cyprus' relates to the southern part of the Island. There is no single authority representing both Turkish and Greek Cypriot people on the Island. Turkey recognizes the Turkish Republic of Northern Cyprus (TRNC). Until a lasting and equitable solution is found within the context of the United Nations, Turkey shall preserve its position concerning the 'Cyprus issue'."

   The following note is included at the request of all the European Union Member States of the OECD and the European Commission: "The Republic of Cyprus is recognized by all members of the United Nations with the exception of Turkey. The information in this document relates to the area under the effective control of the Government of the Republic of Cyprus."

3. These are average values among all survey participants, including those who reported not doing these activities.

References

Australian Communications and Media Authority (ACMA) (2009), *Click and Connect: Young Australians' use of online social media. 02: Quantitative research report, www.acma.gov.au/webwr/aba/about/recruitment/click_and_connect-02_quantitative_report.pdf.*

Austin, K. A., W. D. Lawson and E. Holder (2007), "Efficacy and performance in professional development higher education-sponsored ITV instruction", *Journal of Computing in Higher Education*, Vol. 18(2), pp. 51-81.

Beantin Webbkommunikation (2010), *Internet usage and young Swedes in Sweden, http://beantin.se/post/616872465/internet-use-sweden-young-swedes-children-age-group.*

Drotner, K. (2000), Difference and diversity: Trends in young Danes' media use. *Media, Culture and Society*, Vol. 22(2), pp. 149-166.

European Commission (2008), Flash Eurobarometer (EU27), *Towards a safer use of the Internet for children in the EU – a parents' perspective, http://ec.europa.eu/information_society/activities/sip/docs/eurobarometer/analyticalreport_2008.pdf.*

Ferri, P. *et al.* (2008), *Snack Culture? La dieta digitale degli studenti universitari*, Università Milano Biccoca, Milano.

Hasebrink, U. *et al.* (2011), *Patterns of risk and safety online. In-depth analyses from the EU Kids Online survey of 9-16 year olds and their parents in 25 countries*, LSE EU Kids Online, London.

Helsper, E. and A. Lenhart (2011), Which children are fully online. In L. Haddon and A. Goerzig (eds.), *Children, risk and safety online: Research and policy challenges in comparative perspective*, The Policy Press, Bristol.

Johnson, G. M. and A.J. Howell (2005), "Attitude toward instructional technology following required vs. optional WebCT usage", *Journal of Technology and Teacher Education*, Vol. 13(4), pp. 643-654.

Joiner, R. *et al.* (2007), "The relationship between internet identification, internet anxiety and internet use", *Computers in Human Behavior*, Vol. 23(3), pp. 1408-1420.

Kaiser Family Foundation (2010), *Generation M2, Media in the Lives of 8-Year-Olds*, *www.kff.org/entmedia/upload/8010.pdf.*

Kennedy, G. *et al.* (2006), *First Year Students' Experience with Technology: Are the Really Digital Natives? Preliminary Report of Findings*, Centre for Study of Higher Education, The University of Melbourne, Melbourne.

Kennedy, G. *et al.* (2008), "First year students' experiences with technology: Are they really digital natives?", *Australasian Journal of Educational Technology*, Vol. 24(1), pp. 108-122.

Kennedy, G. *et al.* (2010), "Beyond natives and immigrants: exploring types of net generation students", *Journal of Computer Assisted Learning*, Vol. 26(5), pp. 332-343.

Livingstone, S. *et al.* (2010), *Risks and safety on the internet: The perspective of European children*, LSE, London.

Livingstone, S., K. Ólafsson and E. Staksrud (2011), *Social Networking, Age and Privacy*, LSE EU Kids Online, London.

Media Awareness Network (2005), "Young Canadians in a Wired Word, Phase II; Students Survey", ERIN Research, Canada.

National Center for Education Statistics (2004), *Computer and Internet use. Supplement to 2001 Current Population Survey*, National Center for Education Statistics, Institute of Education Sciences, US Department of Education, Washington, DC.

O'Neill, B., S. Livingstone and S. McLaughlin (2011), *Kids Online: Final recommendations for policy, methodology and research*, LSE EU Kids Online, London.

Organisation for Economic Co-operation and Development (OECD) (2010), *Are the New Millennium Learners Making the Grade? Technology Use and Educational Performance in PISA*, OECD Publishing Paris.

Office of Communications (Ofcom) (2007), *Ofcom's Submission to the Byron Review, Annex 5: The Evidence Base – The views of Children, Young People and Parents, www.ofcom.org.uk/research/telecoms/reports/byron/annex5.pdf.*

Ofcom (2010), *UK children's media literacy, http://stakeholders.ofcom.org.uk/binaries/research/media-literacy/ukchildrensml1.pdf.*

Oliver, B. and V. Goerke (2007), "Australian undergraduates' use and ownership of emerging technologies: Implications and opportunities for creating engaging learning experiences for the net generation", *Australasian Journal of Educational Technology*, Vol. 23(2), pp. 171-186.

Pew Internet and American Life Project (2007), *Teens, Privacy & Online Social Networks. How teens manage their online identities and personal information in the age of MySpace, www.pewinternet.org/~/media//Files/Reports/2007/PIP_Teens_Privacy_SNS_Report_Final.pdf.*

Pew Internet and American Life Project (2009), *Teens and Sexting. How and why minor teens are sending sexually suggestive nude or nearly nude images via text messaging, http://pewInternet.org/Reports/2009/Teens-and-Sexting.aspx.*

Pew Internet and American Life Project (2010), *Reputation Management and Social Media. How people monitor and maintain their identity through search and social media, http://pewinternet.org/Reports/2010/Reputation-Management/Summary-of-Findings.aspx?r=1.*

Rideout, V.J., U.G., Foehr and D.F. Roberts, (2010), *Generation M2: Media in the Lives of 8- to 18-Year Olds,* Henry J. Kaiser Family Foundation, Menlo Park, CA, USA, *http://www.kff.org/entmedia/upload/Generation-M-Media-in-the-Lives-of-8-18-Year-olds-Report.pdf.*

Schmidt, M.E. and E.A. Vandewater (2008), "Media and Attention, Cognition, and School Achievement" *Children and Electronic Media*, Vol. 18(1).

Subrahmanyam, K. *et al.* (2000), "The impact of home computer use on children's activities and development", *Children and Computer Technology* (10), pp. 123-144.

Thinyane, H. (2010), "Are digital natives a world-wide phenomenon? An investigation into South African first year students' use and experience with technology", *Computers & Education*, Vol. 55, pp. 406-414.

Chapter 3

Contrasting views about the digital generation

Doubts emerge around the issue of whether the high level of attachment to digital media and connectivity that can be found among the younger generations is bound to have only positive implications for education, and, consequently, whether schools should follow learners in this respect or, rather contrarily, help them to resist what could be construed as the trivialisation of culture and social interactions. Three main views can be said to dominate the scene: evangelists (or messianists), who promote the idea of a digital generation of learners that will constantly challenge educators; catastrophists, who support the idea that technology attachment is making young people dumb, inattentive, confused and violent; and sceptics, who criticise both evangelists and catastrophists, arguing in the absence of sound evidence the two groups make an ideological option and then present only the evidence that supports their own views. What all this shows is that a solid evidence base is either missing or not widely known.

There are no doubts about the eventual benefits of digital media and connectivity for learning. There are emerging opportunities in distance education at all levels of education, from schools to universities, both in terms of alternative provision arrangements, such as in e-learning or blended learning, and in terms of new forms of content provision, such as open educational resources that connectivity makes potentially available to all. There are also new avenues even for traditional education settings where connectedness can spur innovations in the whole range of processes involved, be they pedagogic or administrative, from lesson planning to learning assessment and feedback, not to mention the ability to use aggregated data for monitoring purposes at the classroom, school and system levels. Finally, there are also increased opportunities for the professional development of teachers and instructors.

The New Millennium Learners and other equivalent concepts are often used as powerful images to evoke two well-known assumptions. The first assumption is that the lives of today's learners are highly dependent on technology to the extent that their social and cultural practices would not exist as they do if digital media were not available anytime, anywhere, to them. The second assumption is that this has important implications for teaching and learning: students are not only accessing, managing, creating and sharing knowledge in dramatically different ways than their teachers often do, but they also have radically new expectations regarding what a quality learning experience should be.

Because the image of the New Millennium Learners is so suggestive and powerful, it is often referred to as an important driver for educational change in education. But in order to evaluate the real magnitude of this phenomenon, it is worth raising a number of questions regarding the empirical validity of the most commonly assumed views.

An old debate with a millennial twist

The public debate about the impact of technological change on media and its implications for children and young people as well as for education is far from being new. One of the most important occurrences was the advent of television. Already in the 1950s, scholars and thinkers split into two opposed visions: a positive perspective (Johnson, 2005) reflected the fascination with the potential of television as a means to spread knowledge, reduce literacy gaps or create broader common social grounds; on the negative side (Popper *et al.*, 1995) concerns were expressed about the risks for democracy, the massive power to manipulate audiences held by relatively few people or the danger of cultural flattening. Critical theories such as the hypodermic needle model (Berger, 1995) focused on all possible dangers. At the same time, enthusiastic early advocates of television as an educational tool started to urge huge investments to make television programmes available for schools use, quite often without a clear understanding of the needs and limitations of school education (Soulages, 2003). As time has passed, it is worth recalling that research has mostly been supportive of a common sense approach, which is "between apocalypse and integration" (Eco, 1973): in the end, it is not technology that counts but how it is used.

The reference to the turn of the millennium also has important resonances. The predicted impact of digital technologies on society and culture has been strengthened by the fact that, incidentally, personal computers became popular in the late 1980s and the Internet and email became a useful commodity for individuals precisely by the end of the millennium. In fact, the concept of "millennium" presses instinctually upon public conscience in a very persuasive way: it is the so-called millenarianism effect, namely the idea that something completely revolutionary could happen at the turning point from one millennium to the next one.

When people think of the turn of a millennium there are often two different reactions: an optimistic one, or a pessimistic one. The reference to the turn of the millennium in the Middle Ages is worthwhile here as two opposing beliefs were diffused: a first and optimistic vision supported the idea of the progress of human society, strengthened by the growing diffusion of great inventions (with a consistent impact on everyday life) like the plough in agriculture or the cheque in economy; a second belief, clearly pessimistic, focused on the loss of the sense of justice and ethics which would lead to the end of the world by the year 1000. The millennial aspect of the public debate on the implications of digital technology on education can easily be seen reflected in the titles of two of the most quoted essays: *The Great Next Generation* (Howe and Strauss, 2000) vs. *The Dumbest Generation* (Bauerlein, 2009). Both titles are in fact a declaration of principle: they synthesise a particular way to explore the unforeseen consequences of technology attachment.

There are, however, important novelties in the current debates in this domain. They are mostly related with the path of technological change, and the uncertainties this brings along, and the nature of digital technology itself.

No matter how familiar people are with digital technologies today, the fact is that they are a relatively recent phenomenon. The first personal computer was launched by IBM in 1981 and the World Wide Web was developed only in 1989. As Box 3.1 shows, digital technologies are an extremely recent phenomenon in human history.

In comparison to non-digital technologies and media, digital media and connectedness provide room for decentralised and uncontrolled interactions among users, proactivity, sharing dynamics, instantaneous feedback effects, synchronous exchanges of knowledge, and more democratic streaming processes. All in all, a whole set of challenging novelties for education (Rivoltella, 2006). Since the widespread advent of mobile devices, this great change in the communication and information field became also ubiquitous and no more limited in timing boundaries. As a result, the perception of a social, cultural and potentially educational revolution is far stronger nowadays than it was when television became a home appliance.

Box 3.1. **Digital technologies and the history of humanity**

Derrick de Kerckhove (2003, 2006) has effectively shown how recent digital technologies are in the context of human development. However, the fact that they have permeated such important spheres of daily life so quickly might suggest that life has always been counting on digital devices. In short:

- 1 700 generations ago, modern man emerges and starts developing language.
- 300 generations ago, he develops writing.
- 35 generations ago, he develops printing.

And then…

- 1910: telegraph and photography.
- 1925: telephone and silent films.
- 1940: radio and talking films.
- 1955: television and mass media.
- 1970: fax.
- 1985: PC and networking.
- 2000: always on the web.

Constructing a social image of connected youth

No matter what you call them – digital natives[1] new millennials[2] or New Millennium Learners – the first generations to grow up surrounded by digital media have reached the age of graduation in many OECD countries. Most, if not all, of them carry a cellular phone and an mp3 player, have a personal computer connected to the Internet, and spend more time in front of its screen than watching television or reading books. This is why the new generations of students, irrespective of age, educational level or programme are also often referred to as the "Net Generation" (Oblinger & Oblinger, 2005; Tapscott, 1999, 2008), the "IM Generation" (which stands for Instant-Message Generation)

(Lenhart, Rainie and Lewis, 2001), the "Gamer Generation" (Carstens and Beck, 2005), for the obvious reference to video games, or even "homo zappiens" (Veen, 2003) for their ability to control simultaneously different sources of digital information and to multitask.

Of course, it is not the first time in history when generational differences have been enhanced in some way by the emergence of new technologies such as television. But there is one big difference between digital technologies and the previous ones: they are personal and, as such, they transform the way in which we communicate and actively manage information and knowledge into a more interactive and instantaneous process, and create an expectation of ubiquitous access to networks.

Drawing on this assumption, media and technology firms tend to exploit the image[3] according to which today's students could be expected to be adept with computers, creative with technology and, above all, highly skilled at multitasking in a world where ubiquitous connections are taken for granted. Some authors have gone even further and claimed that today's students prefer receiving information quickly; are adept at processing information rapidly, in a non-linear way and often while multitasking; have a low tolerance for lectures, preferring instead active rather than passive learning; and rely heavily on communication technologies to carry out social and professional interactions (Frand, 2000; Oblinger and Oblinger, 2005). Going even further than this, it is often assumed that the New Millennium Learners represent a generational phenomenon, with a common set of characteristics, including being sheltered, confident, team-oriented, achieving, pressured, conventional, driven to success, social and experiential learners, and above all multitaskers (Junco and Mastrodicasa, 2008). The idealisation of who a New Millennium Learner is appears depicted in Box 3.2.

What inevitably follows from such assumptions, as some critics have pointed out (Bennett, Maton and Kervin, 2008; Kennedy *et al.*, 2010), is a question about the ability of teachers to deal with these new students' expectations, which some have claimed to be "the biggest single problem facing education today" (Prensky, 2001a, p. 2).

The rationale for such an alarming call can be summarised as follows:

- people born after 1980 have been adopting digital technologies and becoming constantly connected;
- their socialisation and their relation to knowledge and communication takes place in a context permeated by digital media;
- their everyday familiarity with digital media has an impact not only on their technology-related skills but, more importantly, on their social and learning skills;

- this impact makes their expectations regarding teaching and learning radically different from previous generations; and
- therefore, education systems and educators should change to accommodate these new expectations and provide engaging learning experiences.

Box 3.2. **A stereotypical account of a day in the life of a New Millennium Learner (NML)**

It is relatively easy to depict how the stereotyped NML would look in an OECD country. The prevalent stereotype suggests an image according to which a NML wakes up with his cellular phone, playing a song either downloaded or sent to him by another NML. While he switches off the alarm, he will pay a quick look at any incoming messages, and respond if it is worth it. Immediately afterwards, he will inspect the latest news in his inbox or in the social spaces to which he is subscribed, just for a couple of minutes, while opening a music application to hear downloaded tunes while he takes a shower and gets dressed.

During breakfast, if not before, a few more minutes will be devoted to checking any more incoming sms, to see where to gather with friends, for instance. The journey to school will be filled with music as well, by means of an mp player or a multipurpose cellular phone. Then, in most countries, school activities will hardly entail anything but a marginal use of digital technologies and media; therefore, school time may well represent a break in the digitally supported life of the NML – even more since the use of digital personal devices may well be forbidden.

Back home, the computer connected to the Internet will become the epicentre of the activities carried out by the NML, both related to homework and to leisure and social communication. For the former, he will use mostly a word processor and a browser to conduct searches, and to copy and paste relevant information. For the latter, he will use a wide range of web-based applications and sites intended to foster social communication and keep updated with friends irrespective of how nearby they are. While in the case of school-related activities, he will be mostly a passive consumer of already published, posted or shared material, the way in which he will deal with his own sites is far more active, sometimes even producing or uploading his own digital material in the form of mostly pictures, texts, and videos.

With the cellular phone always at hand, he will enjoy dinner in company of the family but it is likely that he will come back to his bedroom to stay connected, watch a movie, play video games and continue posting until rather late in the evening. Both his cellular phone and the computer will remain on overnight.

Yet, doubts emerge around the issue of whether the high level of technology attachment and connectedness that can be found among the younger generations is bound to have only positive implications for education, and, consequently, whether schools should follow learners in this respect or whether they ought to help them to resist what could be construed as the trivialisation of culture and social interactions. For some, this is in itself a window of opportunity for educational change, while for others connectedness is nothing but a gateway to inanity.

Alternative views: evangelists, catastrophists and sceptics

What all this shows is that there is still significant confusion among researchers: a solid evidence base is missing and many contradictory voices populate the debate. Three main views can be said to dominate the scene:

- *Evangelists* (or *messianists*), who accept, believe and promote the idea that a digital generation of learners is good news for education as they will drive a much needed change. Such a generation, thanks to their level of attachment to technology and connectedness would easily become better learners, faster communicators and effective information gatherers, thus leading the way to the knowledge society. It follows from this that educators would be constantly challenged by this digital generation and are likely to be lagging behind learners because of their inferior mastery of technology.
- *Catastrophists*, who also support the idea of a generation-wide phenomenon that challenges educators and traditional education, but believe it will have a negative effect. In short, technology attachment and connectedness will make young people dumb, inattentive, confused, and violent. Socially and culturally speaking, technology may lead to a disaster and have important implications also for education, displacing what really matters from the curriculum and paving the way only for some sort of *edutainment.*
- *Sceptics*, who criticise both evangelists and catastrophists, arguing that neither of them do a good service to education as they are presenting only the evidence that suits their prior ideological assumptions regarding the effects of technology attachment and connectedness for education.

These three alternative views are presented and discussed below.

Evangelism

Digital evangelism in education claims that familiarity with digital media and the relevance that connectedness has in young people's daily life are having an impact on their expectations about learning. Moreover, young people will become the most powerful driver for educational change: they are already asking teachers and schools to adopt an approach to teaching and learning that really suits their needs and expectations. It is generally accepted, for instance, that digital natives demand instant access to information and expect technology to be an integral part of their educational experience (Prensky, 2001a, 2001b; Tapscott, 1999, 2008).

The right educational response, it is said, lies in using technology to put the learner at the centre of the teaching and learning process. This implies in turn that:

a) technology must play a major role in teaching and learning, and

b) that a paradigm shift is required in order to make the most out of technology in education.

Connectedness is considered nothing but good news for education although, as evangelists claim, most teachers are not only reluctant to make use of it but sometimes they are even opposed to this major paradigm shift because of their conservatism: they are still attached to traditional forms of teaching according to which their authority is secured, and they can be in control of learners' access to knowledge. Therefore, lack of attention to the emergence of this new generation of learners causes a number of educational problems, such as a growing tendency towards student disengagement and eventually disaffection, and, consequently, educational failure.

As Rapetti and Cantoni (2010) have shown, digital evangelism may be presented from three different perspectives:

- **Historical**, which emphasises a historical analysis intended to demonstrate the relevance of human interaction with technology for the development of new social practices and skills.
- **Psycho-cognitive**, which goes farther than the historical approach by claiming that the changes in the nature of knowledge are bound to have an impact on human thinking and cognition, also from a psychological perspective.
- **Pedagogical**, which outlines that the changes in learners' approaches to knowledge make them eager to see radical pedagogic changes in schooling.

These three perspectives are succinctly described below.

Using history as an argument to promote digital evangelism

Neil Howe and William Strauss can be taken as the early advocates of this approach in their essay *Generations: the history of America's future, 1584 to 2069* (Howe and Strauss, 1992). Their approach to American history is based on the idea of an endless circle of four archetypes of generations (namely: heroes, artists, prophets, and nomads) succeeding one after another throughout eons. Each generation then develops distinctive characteristics and now it is the turn for Generation Y (referring to the fact that they come after the Generation X) – born in the 1982-2005 period – said to be "heroes".

In their second essay, *Millennials rising: the next great generation*, Howe and Strauss focus precisely on the turn to the third millennium and definitively adopt the expression Millennials. Apparently, this title was chosen after an ABC poll in 1997, in which it received far more support as an expression to describe the young generations than others such as Generation Y, Generation Why, Generation Tech, Generation Next, Gen.com, Generation2000, Echo Boom, Boomer babies, Generation XX, Generation Whatever, Gen-D, Boomlets, Prozac Generation and a host of others. Remarkably, the second most supported was The "Don't Label Us" generation (Howe and Strauss, 2000).

The works of Howe and Strauss have laid down the foundations for the generational approach to dramatic societal changes in particular by suggesting a number of characteristics that can be used to describe current generations. Seven traits emerge from their analysis of the historical characteristics attributed to "heroes" that they apply to the specific sociological context of Generation Y. Namely, the current generation can be said to be:

- *Special.* They feel special because of the enormous amount of attention their parents devoted to them in comparison to previous generations, including educational care and economic wealth.
- *Sheltered.* Again, never before had children received so much protection in relation to their health or education.
- *Confident.* Because parents and educators trust them so much, they in turn place a lot of trust on authority.
- *Conventional.* In opposition to Generation X, they prefer to respect rules of conduct, proper dressing and social authority, and not to take risks. Their revolution is not about "changing the world" but about doing their best in order to score better.
- *Team-oriented.* Due to their experience in school and the opportunities given by the Web 2.0.
- *Achieving.* Generally speaking, they score much better than any other generation before in education.

- *Pressured.* They have been coddled to reach the top. In turn, they feel they have to keep up to the promises. Quite often, young Millennials suffer from anxiety or stress because of this pressure.

Drawing on these seminal works, other authors (Junco and Mastrodicasa, 2007, Wilson and Gerber 2008) have built their own approach and elaborated on the corresponding educational implications. All of them take for granted that this historical approach can provide useful insights into the generational differences, in particular when it comes to the opposition between the adult world and childhood. Flagging out a number of possible critical characteristics of the younger generation may be helpful to start the debate about the effects of technology and connectedness on young people. The analysis of generational differences makes sense only as an initial approach but it is far from being largely supported from a sociological perspective.

Analysing the implications of psychological and cognitive changes

The cornerstone of this analytical perspective lies in Mark Prensky's *Digital Natives and digital immigrants, do they really think differently?* (2001) Drawing on the suggestive concept of the "digital native", opposed to adults being "digital migrants" at most, Prensky was first to claim that children born in the age of computers and the Internet would not come to terms easily with pre-digital culture and teaching. Since this first article, the construct of the digital natives is used to infer that the younger generations are likely to be adept with computers, creative with technology and, above all, highly skilled at multitasking in a world where ubiquitous connections are taken for granted. Following this line of thought, digital natives' attachment to technology and connectedness are assumed to have important implications for the development of their intellectual competencies and cognitive skills, to the extent that they do, in fact, really think differently.

What the psycho-cognitive approach claims, as developed further by other authors, is that the use of digital media has a clear impact on social behaviours and cognition (Palfrey and Gasser, 2008). Young people's attachment to technology is inevitably having an impact on their cognitive development as they process information and communicate in different ways from adults and via multiple channels. Similar claims can be found in other essays, reflecting as well on the impact of exposure to digital media on thinking (Tapscott, 1998; 2009). Whether this will be leading to a new species, the Homo zappiens (Veen, 2006), or not remains and interesting but rhetorical question.

As a matter of fact, digital natives' inevitably short attention spans are the reason why Seymour Papert (1993) coined the term "grasshopper mind", for the inclination to leap quickly from one topic to another, sometimes back and forth, instead of lingering over a subject. Such a repeated behaviour will result in children being impatient if sources of information are not instantly at their fingertips and rarely spending long hours thinking about the same thing. As well as changes in attention spans, the implications of an intensive use of technology may also touch on a broad range of cognitive characteristics, from the need for instant responses to the habit of multitasking and the focus on multimedia content, just to mention a few.

In order to address these potentially shattering claims, neuroscientists started to explore how the brain operates when using technology. In particular, what seems to matter for education is the need to develop a specific ability to manage the overload of information and stimuli: on one hand, this is related to the brain plasticity; on the other, it is enhanced by the dynamics imposed by hypertext and by multitasking. Neurological research (Small and Vorgan, 2008) suggests that people who usually process a continuous stream of data have more neurons dedicated to filtering information. The more this process is run, the more brain plasticity might be developed in the younger ages. This evidence has led some neuroscientists to suggest that tech-savvy people have a greater working memory (the ability to store more information in the short term), and adjust their perception of the world in response to quickly changing information. Because of this, children and young people exposed to digital media are able to "make snap decisions" and "juggle multiple sensors of input", *i.e.* multitasking.

Advocates of this form of digital evangelism suggest that the differences between digital natives and immigrants can have major implications in social life, everyday behaviour, and learning processes, as Prensky remarked in the essay, *Mum, don't bother me, I'm learning* (Prensky, 2006). Thus, the concept of the digital native allows one to elaborate easily on the problems that adults have at addressing the new demands and expectations of younger generations. Inevitably, adults and, in particular parents and teachers, are depicted as digital migrants, struggling to survive in a newfound land owned by digital natives.

Drawing on pedagogical claims

Finally, the pedagogical approach focuses on the following paradox: "technology is everywhere, except at schools". This paradox suggests that children live surrounded by digital media and, irrespective of whether or not technology attachment and connectedness lead to neurological change, the truth is that they are increasingly used to deal with information and communication through digital media. An example of this approach can be found in *Educating the netGeneration* (Oblinger and Oblinger, 2005) as well as in the different works written by Tapscott (1993, 1998, 2009).

How far lessons for educational practice can be drawn from this is yet another issue. As it was pointed out some years ago (Oblinger and Oblinger, 2005): "It is an almost instinctive assumption to believe that Net Gen students will want to use IT heavily in their education; they certainly do in their personal lives. However, if you ask Net Gen learners what technology they use, you will often get a blank stare. They don't think in terms of technology; they think in terms of the activity technology enables. [...] Students often use the word "talk" when they describe text messaging or instant messaging. Software blends into the background."

Advocates of the pedagogical approach make the assumption that young learners, because of their attachment to digital media and connectedness, could be a powerful driver of educational innovation, if only they were allowed to raise their voices. Certainly, there is a point in affirming that if learners master digital media the result is a richer soil for technology-based school innovations. But going further than this, the supporters of the pedagogical approach to digital evangelism foresee a digital school (Ferri, 2008) or university (Junco and Mastrodicasa, 2007) as the only view to respond to the expectations of educational change brought about by today's young learners. Even more radically, it has been also suggested that digital media are a window of opportunity for true children-driven learning, forecasting that "in ten years we will face a completely different school" (Mitra *et al.*, 2005, 21 (3), 407-426; Mitra, 2010).

Pros and cons of evangelism

This optimistic and even enthusiastic view about the effects of digital technologies on learners is having an important influence on public opinion. Its advocates are often quoted by media as relevant sources of wisdom in an area where both parents and teachers seem to be desperately in need of guidance.

However, these optimistic views have also been criticised. It has been often said that they draw only on very peculiar aspects of the social behaviour of adolescents and focus mostly on very impactful characteristics, thus creating a suggestive image – and rather intriguing for adults. Evangelism seems also to have a tendency to generalise observations from non-representative samples to an entire generation (Pletka, 2007) and depict young people as if only digital media mattered to them. Finally, bold statements are made in view of forecasting positive effects on the economy and society thanks to the set of innovative skills that young people develop through their interaction with digital media and connectedness (Veen 2006).

Despite these criticisms, evangelism has also played an important role in pointing to the need to reflect on both the opportunities and challenges brought to education by digital media, and presenting learners as a very important driving force for educational change. In doing this, evangelists have been also instrumental in raising new research domains such as the use of video games in education, the relevance of informal learning through digital media, or the new digital literacies.

Catastrophism

Digital catastrophism in education draws totally opposed conclusions from the same facts marshalled by digital evangelism. For catastrophists, the levels of connectedness are alarming and, if not controlled, will erode the foundations of good education. Their concern stems from the amount of time devoted to what they see as useless, if not pernicious, activities online that could have been invested instead in serious and formative activities. Such a concern goes beyond the risks of technology attachment and claims the loss of appreciation for academic content and for the skills and values that have traditionally contributed to a sound classical education.

Therefore advocates of catastrophism argue that connectedness is harming education. They clearly support the principle that schools should teach students to appreciate the benefits of traditional, classical culture. For instance, in *The dumbest generation. How the Digital Age Stupefies Young Americans and Jeopardizes Our Future* (Bauerlein [2009] claims that technology stupefies the younger generation by contracting their horizons to only themselves and offering them easy detours away from practicing solid intellectual skills. Teachers are seen as the last bastion of resistance against detrimental influences that could culminate the younger generations being totally lost. For example, it has been stated that the younger are "hyper-networked kids who can track each other's every move with ease, but are largely ignorant of history, economics, culture, and other subjects… For digital immigrants, people who are 40 years old who spent their college time in the library acquiring information, the Internet is really a miraculous source of knowledge. Digital natives, however, go to the Internet not to store knowledge in their minds, but to retrieve material and pass it along. The Internet is just a delivery system…" (Bauerlein, 2009, p. 40).

Under a catastrophist perspective young people are said to be:

- Dumb and ignorant (Bauerlein, 2009);
- Violent, online bullies and net addicted to pointless activities (Mothers Against Video games Addiction and Violence, 2002-06);
- Shameless (Durham, 2008);
- Coddled, adrift and slackers (Damon, 2008);

- Narcissists (Twenge *et al.*, 2008);
- The ones who take Google as Gospel (Keen, 2007); and
- The ones who click, instead of think, when looking for knowledge (Brabazon, 2002),

Catastrophism can also become even more frightening by way of scaling up the dangers of connectedness, that is, by assuming that connectedness *per se* leads to risky behaviours. Issues related to sexual harassment, paedo-pornography, net-addiction, voyeurism, exhibitionism, and anorexia or bulimia are usually evoked and documented. Clearly these risks exist but connectedness can hardly be seen as their main cause. Rather, what all this reflects is the need for revisiting parenting and education in a context where digital media plays such an important role in the lives of children and young people.

Scepticism

At first glance it may seem that evangelism and catastrophism reflect alternative interpretations of the same phenomenon. Their respective conclusions and recommendations for action are driven by alternative views of education. While the evangelists take the whole issue of technology attachment and connectedness as an opportunity to assert the need for educational change, the catastrophists hold that a good quality education must insist on traditional values, content and methods and use technology to contribute to supporting them. Evangelists and catastrophists can be said to draw different conclusions from the same phenomenon.

A deeper analysis shows, however, that the knowledge base used by each perspective is far from being the same. Evangelists and catastrophists construct a different stereotyped image of the impact of technology attachment, and particularly of connectedness, on learners' views and expectations. As a matter of fact, the common base is simply reduced to statistical data showing that the level of pervasiveness of technology among young people is much higher in comparison to that of older generations. Both evangelists and catastrophists infer different conclusions and implications from this, following their own views on what a good education should be. In the end, each perspective elaborates a different image of youth, and their respective educational recommendations follow accordingly from either an optimistic or pessimistic assessment of that image.

Scepticism may be seen as a sensible scientific attitude towards new or emerging phenomena for which no previous knowledge exists. A sceptical attitude is particularly appropriate as a starting point for research in the social sciences where claims can be easily made using weak evidence. With this attitude a number of academics started recently to challenge the well-known assumptions made by both evangelists and catastrophists. Some of the sceptics have argued the need for an evidence-based analysis (Bennett, Maton and Kervin, 2008) while others have made an effort to critically review the lack of evidence of simplistic approaches. As Bullen *et al.* (2009) have said: "a review of literature on the millennial learner and implications for education reveals that most of the claims are supported by reference to a relatively small number of publications. The works most often cited are Oblinger and Oblinger (2005), Tapscott (1998), Prensky (2001a, b), and Howe and Strauss (2000). Other works that are often mentioned, although less frequently, include Seely-Brown (2002), Frand (2000) and Turkle (1995). What all of these works have in common is that they make grand claims about the difference between the millennial generation and all previous generations and they argue that this difference has huge implications for education. But most significantly, these claims are made with reference to almost no empirical data. For the most part, they rely on anecdotal observations or speculation. In the rare cases, where there is hard data, it is usually not representative."

There are still significant questions surrounding the main claims made about a new generation of learners and, for the most part, the debate has been neither empirically nor theoretically informed (Jones *et al.*, 2010). For instance according to Bennet *et al.* (Bennett and Maton, 2010; Bennett *et al.*, 2008) the debate can be linked to an academic form of "moral panic" in which arguments and assertions are couched in overly dramatic language which then lead to appeals for urgent action and fundamental change. They also note that this academic moral panic is associated with polarised and determinist arguments in which descriptions rely on oppositions between digital natives and digital immigrants, between a new Net generation and all previous generations, and is one in which change is portrayed as inevitable, with those who resist these characterisations being represented as simply resistant to change, out of touch with reality and not having legitimate concerns. This critique of the language of moral panic suggests that it closes down debate and allows unsupported claims to circulate and gain credibility. In the end what happens is that the arguments about the pressures arising in education from a new generation of New Millennium Learners lead to a one-way determinism forcing institutions and teachers to change.

There is also a paradox lying at the centre of this debate because each person is fixed by their own generational position: one either is or is not a "native". But when one is not, what emerges implicitly is that one should be like a digital native, because then one would be more equipped to tackle the challenges and reap the benefits of the knowledge economy and society. So a requirement to change and become more like digital natives is transformed into a professional imperative for teachers. In turn, this leads to a deficit model for professional development in which, however hard older teachers try, they will never be able to bridge the gap arising from their generational position. Moreover, it has also been argued that the persistence of this image of the Net generation or digital native student has a strong relationship to the logic of the market and a culture of enterprise, which is evident in advertising where these ideas are clearly reinforced by marketing aimed at the education sector (Bayne and Ross, 2007).

Both evangelists and catastrophists tend to make universal statements about a generation that they assume to be totally homogeneous in relation to digital media attachment and skills. Yet, "a bunch of people" is not a generation. Moreover, the only digital media skill that is definitely attributable to young people (who are used to digital media) is the information gathering ability (Bayne and Ross, 2007; Schulmeister, 2008).

Gross generalisations and stereotypes can be handy but in education they are often misleading. Evangelists and catastrophists coincide to being over-deterministic, because they describe people, their behaviours, habits and beliefs, and make claims about how they will learn and behave in the future, simply drawing on the fact that most young people are growing up in a world permeated by technologies (Rapetti, Cantoni and Misic, 2009). More important than a general attribution is the link between particular levels of intensity of use and the actual activities performed with digital media and connectedness with learning preferences and outcomes (Gamero, 2010; Selwyn, 2009; Sigman, 2008; Kolikant, 2010).

In addition, it cannot be ignored that the most influential works come from North America and that there are clear risks in applying exactly the same analysis to other countries (Bullen, Tannis and Qayyum, 2008-present). The same applies to the educational implications derived from these analyses, which cannot be said to be easily extended to all possible educational contexts (Bullen *et al.*, 2009).

Finally, the deterministic attitude towards the generational gap may lead to an "educational give-up" (Bennett, Maton and Kervin, 2008), provoked by the moral panic of no longer being able to communicate with someone completely different and much more knowledgeable (Bayne and Ross, 2007).

Is there room for yet another approach?

The appealing idea of a generational difference in approaches to learning due to a massive exposure to technology has been the starting point for two decades of a complicated, nebulous and controversial debate. As such, it has been fed by a corpus of information that, if approached globally, gives the impression of a very scattered field, with too many, too different and often inconsistent contributions. As a result, a useful and clear knowledge base is missing and without it guidance for parents, educators and policy makers is often based on intuitions.

What the evidence says

A nuanced understanding of the extent and nature of technology attachment and use of connectedness by young people requires a deep insight into the contexts in which the technologies are being used: the subject of a course; its pedagogic design; students' socio-economic background; their life circumstances, such as affluence, geographic proximity to friends and family; and personal, psychological characteristics, such as sociability and openness to new experiences (Schulmeister, 2008).

A range of empirical studies investigating students' use of technologies has been published in recent years. These empirical studies, conducted in different countries worldwide and in different types of institutions, are reaching very similar conclusions suggesting that the "digital native" label may be too simplistic to explain the ways young people use technologies.[4] Box 3.3 presents a few examples of this type of research.

Box 3.3. **Challenging the universal nature of digital natives**

Despite the claims of many analysts, empirical research conducted in a large variety of countries tends to show that a much more nuanced perspective is needed. As it happens with learning styles, engagement with technology and attitudes towards its use in formal learning environments varies greatly depending on a number of individual variables.

In Australia, Kennedy *et al.*, 2008 found lack of homogeneity in technology adoption patterns, particularly when moving beyond established technologies such as mobile phones and email. They concluded that "the widespread revision of curricula to accommodate the so-called Digital Natives does not seem warranted" because "we cannot assume that being a member of the Net Generation is synonymous with knowing how to employ technology strategically to optimise learning experience in university settings."

In Austria, Nagler and Ebner (2009) found an almost ubiquitous use of Wikipedia, YouTube and social networking sites while social bookmarking, photo sharing and microblogging were much less popular and concluded that "the so-called Net Generation exists if we think in terms of basic communication tools like e mail or instant messaging. Writing an e mail, participating in different chat rooms or contributing to a discussion forum is part of a student's everyday life" (2009, p. 7).

In Canada, Bullen *et al.*, 2008 investigated students' fit to the "millennial" profile. Their results suggest that students "use a limited toolkit" (2008, p. 8) and that the adoption of these tools was driven by the familiarity, cost and immediacy considerations. Their findings

Box 3.3. **Challenging the universal nature of digital natives** *(continued)*

show that while there was no evidence to suggest that students have a deep knowledge of technology, they use technology in very context sensitive ways. Within an identified set of tools, students were able to identify which was better suited to a given task. There was little evidence to support a claim that digital literacy, connectedness, a need for immediacy, and a preference for experiential learning were characteristics of a particular generation of learners.

In the United Kingdom, a 2007 survey by Synovate found that only 27% of British teenagers could really be described as having the kind of deep interest and facility in technology that the label of digital natives implies. The majority (57%) use relatively low level technology to support their basic communication or entertainment needs and there is a substantial residuum of 20% ("digital dissidents") who actively dislike technology and avoid using it wherever possible (Williams and Rowlands, 2008). Jones *et al.* explored undergraduate students' access to hardware and Internet, and their use of digital technologies in learning and leisure activities. Authors suggest that "the idea that the Net generation are more likely to be inclined to participation [on the web] may be somewhat exaggerated" (Jones *et al.*, 2010). The study recorded low levels of usage of blogs, wikis and, particularly, virtual worlds. When asked about their confidence in using specific technologies, the majority of students reported slight and basic confidence in using established technologies such as presentation software, online library resources and spreadsheets. Over a third reported no confidence or minimal skills in using conventional learning platforms, writing and commenting on blogs or using wikis. The authors concluded that "it does not seem that [students] are marked by their exposure to digital technologies from an early age in ways that make them a single and coherent group…" (Jones *et al.*, 2010, p. 19). Policy makers should be cautious against adopting technological determinist arguments that suggest that educational institutions simply have to adapt to a changing student population who are described as a single group with definite and known characteristics.

In the United States, Hargittai (2010) conducted a quantitative study of undergraduate students' Internet use. The study focused on the role of "context" – socio-economic status, self-reported skills, experience and autonomy in technology use – in bringing about differentiated use of technology. The study found a "considerable variation, even among fully wired college students when it comes to understanding various aspects of Internet use. These differences are not randomly distributed. Students of lower socio-economic status, women, students of Hispanic origin, and African Americans exhibit lower levels of web know-how than others" and suggested that "differentiated contexts of uses and experiences may explain these variations" (2010, p. 108). Data simply does not support the premise that young people are universally knowledgeable about the web.

Moving forward

Therefore, although the arguments sustaining the existence of digital natives as an overwhelming generational change have been well publicised and uncritically accepted by some, there is no empirical basis to support this contention. Counter-positions have recently emerged which emphasise the need for robust evidence to substantiate the debate and to provide an accurate portrayal of technology adoption among students (Bennett *et al.*, 2008; Margaryan, Littlejohn and Vojt, 2010; Schulmeister, 2008; Selwyn, 2009).

Probably these initial approaches were well received, gained prestige and soon influenced the public attitude towards the stereotype of the connected youth for a number of reasons: it explained cogently and briefly a complex reality, which was worrying for parents and educators; it offered a common sense-based interpretation, which was easy to understand as it was based upon the everlasting contraposition between the younger and the older generations; and finally it filled an important gap of knowledge in education.

On the other hand, it may well be that, on the whole, digital technologies are too recent, and their effects on learners too multi-faceted and interrelated – and hence difficult to untangle – to allow the research community to provide a quick and coherent knowledge base for the concerned stakeholders, thus competing with these initial approaches. Nevertheless, the time has come to explore this scattered field. And, in fact, in a number of OECD countries there is already an emergent body of research, from a variety of disciplines, which, if added up and compared, may well contribute to clarifying the issues at stake and, further, supporting and informing with evidence the policy debate.

More empirical research is certainly needed to improve our understanding of the nature and extent of technology uptake by students. In parallel to understanding what tools students use and how they use them, it is also important to elucidate the role of digital technologies in students' learning, because "it is not technologies, but educational purposes and pedagogy that must provide the lead, with students understanding not only how to work with ICTs, but why it is of benefit for them to do so" (Kirkwood and Price, 2005, p. 275). There is a need therefore for a consistent knowledge base about how connectedness influences learners and their expectations about teaching and learning. Such a knowledge base will immediately result in a much more nuanced view of the issues around connectedness and education.

Yet, a solid knowledge base is not enough, neither for educators nor for policy makers in education. Some considerations must be taken into account. Firstly, it cannot be assumed that the effects are going to be stable over time. Contrarily, given the rapidly evolving nature of technological and social change in this domain, the analysis of the effects of technology attachment and connectedness on learners has to be constantly updated. Secondly, although the knowledge base can greatly improve the public debate about this issue, by nurturing it with hard evidence, anecdotal evidence and tacit knowledge should not be underestimated, particularly when educators face for their first time challenging student behaviours or demands.

Notes

1. A term coined by Mark Prensky (2001a) in his much quoted essay "Digital natives, digital immigrants". As he claims, "digital natives" are native speakers of the digital language of computers, video games, and the Internet, as opposed to their teachers who are mostly "digital immigrants" and have had to adapt to the new environment created by technology. They have had to learn a new language and, because of this, their accents in speaking this digital language are still discernable.
2. This other term was first used by generational historians and sociologists Howe and Strauss (2000) in their essay entitled "Millennials rising: the next great generation" to describe what they thought to be quite a different generation from the previous one, the so-called Generation X.
3. The number of articles in newspapers contributing to elaborate this hype is countless. A good, well-supported example can be found in the piece by Zeller, Bernstein and Marshall (2006).
4. While these studies provide a valuable and much needed contribution to the body of empirical evidence in this area, they share a number of limitations, and further methodological refinements are needed as Margaryan *et al.* (2010) have indicated.

References

Bauerlein, M. (2008), *The Dumbest Generation: How the Digital Age Stupefies Young Americans and Jeopardizes Our Future (Or, Don't Trust Anyone Under 30)*, Tarcher/Penguin, New York.

Bayne, S. and J. Ross (2007), "The 'digital native' and 'digital immigrant': A dangerous opposition", paper presented at the Society for Research into Higher Education (SRHE), *www.malts.ed.ac.uk/staff/sian/natives_final.pdf*.

Bennett, S. and Maton, K. (2010), "Beyond the 'digital natives' debate. Towards a more nuanced understanding of students' technology experiences", *Journal of Computer Assisted Learning,* Vol. 26(5), pp. 321-331.

Bennett, S., K. Maton and L. Kervin (2008), "The 'digital natives' debate: a critical review of the evidence", *British Journal of Educational Technology*, Vol. 39(5), pp. 775-786.

Berger, A. (1995), *Essentials of Mass Communication Theory*, SAGE Publications, London.

Brabazon, T. (2002), *Digital hemlock: Internet education and the poisoning of teaching*, UNSW Press, Sydney.

Bullen, M. *et al.* (2009), "The Net Generation in Higher Education: Rhetoric and Reality", *International Journal of Excellence in e-Learning,* Vol. 2, No. 1

Bullen, M. *et al.* (2009), "The Net Generation in Higher Education: Rhetoric and Reality", *International Journal of Excellence in e-Learning*, Vol. 2, No. 1.

Bullen, M., M. Tannis and A. Qayyum, (2012), *The net gen skeptic blog, www.netgenskeptic.com/* accessed since 2008 to present date.

Carstens, A. and J. Beck (2005), "Get Ready for the Gamer Generation", *TechTrends*, Vol. 49(3), pp. 22-25.

Damon, W. (2008), *The path to purpose: helping our children find their calling in life*, Free Press, New York.

Durham, M. and Gigi (2008), *The Lolita Effect,* Overlook Press, New York.

Eco, U. (1973), *Apocalittici e integrati,* Bompiani, Milano.

Ferri, P.(2008), *La scuola digitale come le nuove tecnologie cambiano la formazione,* Mondadori, Milano.

Frand, J.L. (2000), "The information-age mindset. Changes in students and implications for higher education", *Educause review*, Vol. 35(5), pp. 15-24.

Hargittai, E. (2010), "Digital Na(t)ives? Variation in Internet Skills and Uses among Members of the Net Generation", *Sociological Inquiry,* Vol. 80(1), pp. 92-113.

Howe, N. and W. Strauss (1992), *Generations: The History of America's Future, 1584 to 2069*, Quill, New York.

Howe, N. and W. Strauss (2000), *Millennials Rising: The Next Great Generation*, Vintage Original, New York.

Johnson, S. (2005), *Everything Bad Is Good for You: How Today's Popular Culture Is Actually Making Us Smarter,* Riverhead Books, New York.

Jones, C., *et al.* (2010), "Net generation or Digital Natives: Is there a distinct new generation entering university?", *Computers and Education*, Vol. 54, pp. 19.

Junco, R. and J. Mastrodicasa (2007), *Connecting to the net.generation: what higher education professionals need to know about today's students*, NASPA, Student Affairs Administrators in Higher Education, Washington, DC, *http://blog.reyjunco.com/pdf/NetGenerationProof.pdf.*

Keen, A. (2007), *The cult of the amateur: how today's internet is killing our culture*, Doubleday Currency, New York.

Kennedy, G. *et al.* (2008), "First year students' experiences with technology: Are they really digital natives?" *Australasian Journal of Educational Technology,* Vol. 24(1).

Kennedy, G. *et al.* (2010), "Beyond natives and immigrants: exploring types of net generation students", *Journal of Computer Assisted Learning*, Vol. 26(5), pp. 332-343.

Kirkwood, A. and L. Price (2005), "Learners and learning in the twenty-first century: what do we know about students' attitudes towards and experiences of information and communication technologies that will help us design courses?" *Studies in Higher Education*, Vol. 30(3), pp. 257-274.

Lenhart, A., L. Rainie and O. Lewis (2001), *Teenage Life Online: The Rise of Instant-Message Generation and the Internet's Impact on Friendship and Family Relationships*, Pew Internet and American Life Project, Washington, DC.

Margaryan, A., A. Littlejohn and G. Vojt (2010), "Are digital natives a myth or reality? University students' use of digital technologies", *Computers & Education*, Vol. 54(3), April 2010, pp. 722-732.

Mitra, S. *et al.*, (2005), "Acquisition of computing literacy on shared public computers: Children and the 'hole in the wall'", *Australasian Journal of Educational Technology*, Vol. 21, No. 3, pp. 407-426, *www.ascilite.org.au/ajet/ajet21/mitra.html.*

Mothers Against Video games Addiction and Violence, (MAVAV) (2002-2006), *www.mavav.org/resources/*

Nagler, W. and M. Ebner (2009), *Is your university ready for the Ne(x)t-Generation?*, paper presented at the Proceedings of 21st world conference on educational multimedia, hypermedia and telecommunications (EDMEDIA), Honolulu.

Oblinger D. and J. Oblinger (2005), *Educating the Net Generation*, EDUCASE, *www.educase.edu/educatingthenetgen/*

Palfrey, J. and U. Grasser (2008), *Born digital: understanding the first generation of digital natives,* Basic Books, New York.

Papert, S. (1993), *The children's machine: rethinking school in the age of the computer*, Basic Books, New York.

Pletka, B. (2007), *Educating the net generation: how to engage students in the 21st century*, Santa Monica Press, Santa Monica, CA.

Popper, K. *et al.* (1995), *La télévision : un danger pour la démocratie*, Anatolia Editions, Paris.

Prensky, M (2006), *Don't bother me Mom, I'm learning : How computer and video games are preparing your kids for twenty-first century success and how you can help!*, Paragon House, St. Paul, Minn., USA.

Prensky, M. (2001), "Digital Natives, Digital Immigrants, Part II: Do They Really Think Differently?", *On the Horizon,* NCB University Press, Vol. 9 No. 6.

Prensky, M. (2001a), "Digital Natives, Digital Immigrants", *On the Horizon*, Vol. 9(5), pp. 1-12.

Prensky, M. (2001b), "Digital Natives, Digital Immigrants, Part II: Do They Really Think Differently?", *On the Horizon*, Vol. 9(6), pp. 15-24.

Rapetti, E. and L. Cantoni (2010), "Digital natives and learning with the ICTs the 'GenY @ work' research in Ticino, Switzerland", *Journal of E-Learning and Knowledge Society*, Vol.6 no. 1, pp. 39-49,

Rapetti, E., Cantoni, L. and Misic, S. (2009), "New cultural spaces for learning: the learners' voices", *Online Webasculture Conference Proceedings*, July 16-18, Giessen, Germany, *www.webasculture.de/index.php?id=75.*

Rivoltella, P. (2006), "Screen generation : gli adolescenti e le prospettive dell'educazione nell'età dei media digitali", Pedagogia e scienze dell'educazione, Ricerche,Vita and Pensiero, Milan, Italy.

Schulmeister, R. (2008), *Gibt es eine "Net Generation"?*, retrieved 5 March 2010 from *www.zhw.uni-hamburg.de/uploads/schulmeister_net-generation_v3.pdf.*

Selwyn, N. (2009), "The digital native: myth and reality, *Aslib Proceedings: New Information Perspectives*, Vol. 61(4), pp. 364-379.

Small, G. W. and G. Vorgan (2008), *IBrain : surviving the technological alteration of the modern mind*, Collins Living, New York.

Soulages, C. (2003), *Apprendre la télévision, observer, monter, analyser, comprendre le JT,* INA/CEMEA/CLEMI, Paris.

Tapscott, D. (1998), *Growing Up Digital: The Rise of the Net Generation*, McGraw-Hill, New York.

Tapscott, D. (2008), *Grown Up Digital: How the Net Generation Is Changing Your World*, McGraw-Hill, New York.

Tapscott, D. and A. Caston (1993), *Paradigm shift : the new promise of information technology,* McGraw-Hill, New York.

Twenge *et al.* (2008), "Egos Inflating Over Time: A Cross-Temporal Meta-Analysis of the Narcissistic Personality Inventory", *Journal of Personality*, Blackwell Publishing, Vol. 76, No. 4.

Veen, W. (2003), "A New Force for Change: Homo Zappiens", *The Learning Citizen* (7), pp. 5-7.

Veen, W. and B. Vrakking (2006), *Homo Zappiens*, Network Continuum Education, London.

Williams, P. and I. Rowlands (2008), *Information behaviour of the researcher of the future: the literature on young people and their information behaviour*, British Library and JISC, London.

Wilson, M. and L.E. Gerber (2008), "How Generational Theory Can Improve Teaching: Strategies for Working with the 'Millennials'", *Currents in Teaching and Learning*, Vol. *1*(1), pp. 29-44.

Zeller, T., D. Bernstein and C. Marshall (2006), "A Generation Serves Notice: It's a Moving Target", *The New York Times*, 22 January.

Chapter 4

What are the effects of attachment to digital media and connectivity?

This section presents and discusses the main research findings in controversial areas such as the effects of digital technologies and connectivity on cognitive skills development and social values and lifestyles. The results of empirical research, particularly when considering meta-analyses, give the impression of a very scattered field, with only a few efforts made to accumulate knowledge in a way that becomes useful information for parents, educators and policy makers. In addition, the available yet scarce research evidence does not always present a coherent picture, with results from some studies disagreeing with those of others. It may well be that, on the whole, digital technologies are too recent, and their effects on learners too multi-faceted and interrelated – and hence difficult to untangle – to allow the research community to provide a coherent knowledge base to the stakeholders concerned.

It is reasonable to expect that if digital media and connectivity play such an important role in students' lives, this is bound to have an effect on their cognitive skills, lifestyles and social values, as well as the way they look at their own education and the role that technology should play in it (see Chapter 5). The attractiveness of this reasoning is difficult to deny, but does reality confirm and validate these assumptions about the New Millennium Learners?

The central question for researchers is therefore not whether technology is affecting cognitive development because that can be taken as a given. As neuroscience specialists have phrased it (Bavelier, Green and Dye, 2010), the relevant question is instead, how is technology affecting cognitive development? Are the changes for the better or for the worse? How can we harness technology to effect more changes for the better? How do we limit technology's ability to effect changes for the worse? And all these questions have to be addressed bearing in mind that "technology" embraces multiple entities, and thus it is unlikely to have a single unique effect.

Hence the importance of backing up these hypothetical assumptions with hard data and addressing research questions such as the following: are today's students different from previous generations because of their continuous exposure to digital media and their experience of being permanently connected? When today's students enter the classroom, do they also bring new expectations regarding teaching, learning and, particularly, the use of technology?[1] On the surface these seem to be very simple questions to which one could expect straightforward answers,[2] but is there enough research evidence to accurately address them?

A research challenge

In fact, in a number of OECD countries there is already a substantive body of research from a variety of fields, which, if added up and compared, may well contribute to clarifying the issues at stake and, further, supporting and informing the policy debate with evidence. Four messages emerge very clearly:

- There is sustained evidence regarding the effects of technology on cognitive skills development, particularly in areas related to visual-spatial skills and nonverbal forms of intelligence. Despite this, the most potentially interesting domains, at least from an educational point of view, have not been documented by empirical research. Areas such as those related to information processing, reflective and critical thinking, creativity and, meta-cognitive skills in general are very often missing in this regard.

- Insofar as digital technologies are added to an already complex picture of media exposure, the reported effects on socialisation are varied. On the one hand, it has been shown that time devoted to digital technologies adds to time devoted to other media and thus reduces time spent on family interaction or face-to-face peer interaction. But, on the other hand, time devoted to digital technologies gives rise to other forms of socialisation in a third space, virtual by nature, which is less exposed to responsible adult supervision or counseling.
- There is enough empirical evidence to sustain the idea that playing with video games that support violence or sexual stereotypes does have a negative effect on young people, particularly if use is excessive. In addition, socio-economic factors, such as parental education level, are likely to have an impact too (Rideout, Foehr and Roberts, 2010).
- There is an intrinsic difficulty when researching the effects of technology on educational performance. In practice, there is no conclusive evidence about the effects of technology upon academic achievement. This is partly for obvious reasons – digital technologies are tools that can be used with a wide range of methodologies and strategies, for which there is a great deal of empirical research with inconclusive results – but also because the correct methodological approach, which should involve large longitudinal studies, has not been put in place yet.

The following sections discuss each of these messages in more detail.

Cognitive skills development

Do computers and the Internet have any effects on children's cognitive skills and development overall? It is already clear from well-documented research (Bracken and Lombard, 2004) that children can recall what they learn from a computer, especially if they are rewarded. However, the crucial question is how computers are affecting learning the skills necessary for reasoning, problem solving, reading and creativity.

At least potentially, digital media contain features that provide opportunities for enhancing various cognitive skills. Throughout the years, their use in formal and informal settings has been related to training or effects on memory skills, attention, executive functions such planning and strategy use, language, thinking and visual-spatial skills. Research in these areas has taken various directions: understanding how different digital materials are processed; what cognitive abilities they activate and affect; how mental models are built from digital aids; how to train certain cognitive skills in the context of academic learning; and how the use of computers and the Internet at home for playing games, searching for information, and communicating affect cognitive skills.

For example, researchers claim that the exposure to the proliferation of imagery in media has contributed to the selective increases in nonverbal intelligence scores during the past century in industrialised countries, according to the so-called Flynn effect.[3] A comparison of average scores on the Raven Progressive Matrices test (nonverbal) among British adults of comparable ages in 1942 and 1992 showed significant increases for all age groups tested (Greenfield, 1998).

Flynn's current explanation (Flynn, 2007) is that environmental changes arising from modernisation – such as more intellectually demanding work, greater use of technology and smaller families – have meant that people are far more used to manipulating abstract concepts such as hypotheses and categories than a century ago. Substantial portions of IQ tests deal with these abilities.[4] However, other researchers (Neisser, 1998) have found that the Flynn effect may be largely due to increased performance on nonverbal items, especially on items that draw on spatial visualisation. Hence the Flynn effect might also be due to the increased use of audiovisual media by children. Children's exposure to computer screens and, particularly, to video games may have a similar long-term effect, individually and perhaps generationally, and there is already some research documenting it (Newcombe and Huttenlocher, 2000).

However, it should be pointed out that most of the research addressing issues concerning technology and cognition has been qualitative in orientation, addressing issues to do with how technology facilitates reasoning rather than with whether it has discernible effects on cognitive processes and representations. Moreover, quantitative methods are used to study interrelationships between personal/cognitive and situational variables, rather than in order to experimentally isolate the effects of technology on cognition, something that would make the research less ecologically valid (Arnseth and Ludvigsen, 2006). In this respect, the most quoted research review, restricted only to empirical research, concluded that the most crucial question – whether technology enhances child development – has an affirmative answer (Koeppl *et al.*, 2001). Such a positive answer explains why this research review is so often quoted, despite the fact that a closer look reveals that the reviewers paid attention only to empirical research focusing on the beneficial impacts of technology on child development.[5] Needless to say, research focused exclusively on the positive side of technology may be seen to reflect the opinions of some interest and advocacy groups, as Atkinson *et al.* (2001) have already pointed out.[6]

The related notion that the brains of today's learners are different from their teachers' is very controversial, with neurophysiologists and cognitive psychologists currently engaging in debates involving unresolved issues such as working memory differences (Baddeley and Logie, 1999; Niederhauser *et al.*, 2000) or fluid intelligence/spatial ability differences (Ackerman, Beier and Bowen, 2002; J. R. Anderson, 2000). On the other hand, there is clear physiological evidence of substrate uniformity; in other words, the human brain has not changed (Eimer, Van Velzen and Driver, 2002; Shimojo and Shams, 2001; Wright, Fields and Harrison, 2000) – or at least not yet.

The evidence from research on the impact of digital media use on cognitive skills is difficult to generalise as it is always placed in context and set in relation to a number of factors such as age, gender, socio-economic background, time spent in computer activities, and preference for certain activities. However, the majority of inquiries in this regard examine how the intended use of digital technologies in the form of training affects certain skills, competencies and behaviours. Although such evidence might not show how technology use affects learners in broad terms, it improves the understanding of the processes taking place and enables the development of interventions and learning scenarios. Additionally, some of the effects observed in controlled environments can be expected to occur in the everyday interaction of young people with educational software, computer games or the Internet.

With these limitations in mind, the following paragraphs present the existing evidence in areas where it can be said to be conclusive: namely, visual-spatial and, memory skills and, to a lesser extent, the effects of multitasking on learning and cognitive development.

Visual-spatial skills

The ability to deal with two and three dimensional images, spatial visualisation, and the skills needed to read images, the ability to recognise the information which images contain, as well as the ability to interpret images, are expected to improve with repeated practice, for example, through the regular use of multimedia and computer games.

Probably the largest part of the empirical research regarding the impact of digital media use on cognitive skills focuses on the development or training of visual-spatial skills. Many computer applications have design features that require visual rather than verbal information processing. The constant enhancement of the graphic design and realism of video games in the last years presents new dimensions of spatial, iconic and dynamic features, which provide a new environment for children to develop a set of skills concerning visual attention, orientation and spatial representation.

The most often quoted overview of the research on promoting spatial abilities through computer games (Souvignier, 2001) found consistent evidence that such skills can be improved through drilling and practising, as often happens when playing video games. Recent studies outline that playing video games improves visual attention, meaning that heavy users are better in allocating attention and in filtering out irrelevant information (Dye, Green and Bavalier, 2009). The debate is open, since it seems that using electronic visual media may enhance visual attention and visual-spatial processing, but does not necessarily cultivate a better order processing in terms of cognitive skills related to learning field (Greenfield, 2009).

Memory skills

Another frequent topic is the impact of digital media use on memory skills, which has received a great deal of attention in relation to the research on the impact of violent media content on young people. While some structural features of multimedia facilitate retention, it is argued that strong emotional experiences during interaction with digital media might hinder memory skills and the long-term effects of learning. The assumption behind this hypothesis is that emotional events influence how things are perceived and remembered. The major concern is that nowadays computer games and movies are made to provoke strong emotional responses, and that engaging in such activities after learning would decrease the learning effects from school or homework. Mößle *et al.*, (2006) report such a correlation between longer playing times of computer and video games and low academic achievement, based on a survey of 6 000 4th graders in Germany. However, such a correlation does not imply a causal relationship and there could be a host of other variables affecting game playing and/or academic achievement.

A hotly debated point concerns the hypotheses of the growing capability of humans' working memory, thanks to the massive usages of devices providing a huge amount of information to process at the same time. Unfortunately, there is no agreement among scholars about the relationship between attention and working memory; and it is yet an open issue how the decrease in attention provoked by many attending to many stimuli simultaneously affects working memory capability (Klingberg, 2009).

Media multitasking

Before examining the evidence related to the effects of multitasking on young people's learning and skill development, it is useful to distinguish between the different ways researchers have been using the term. A useful way to think of multitasking is in terms of three dimensions: the medium used, the task being carried out and the goal aimed at. This results in the following typology:

- Different media, different tasks, different goals (*e.g.* mobile phone-TV, writing SMS-watching, communication-entertainment);

- Different media, different tasks, same goal (*e.g.* mobile phone-PC, talking-surfing the net, homework);
- Same media, different tasks, same goal (*e.g.* PC, searching-writing, homework);
- Same media, different tasks, different goals (*e.g.* PC, listening to music-chatting, entertainment-communication).

Media multitasking (namely, using more than one device at the same time), the multitasking within the same media (*e.g.* using a smartphone to surf the net while having a phone call) inspired the idea of brain multitasking, that is, the hypothesis that the brain process more than one flow of information simultaneously (Wallis, 2010).

However, the "response selection bottleneck" process limits brain multitasking, as demonstrated in a number research studies (Dux et al, 2006). This means that concurrent tasks, at any given moment, oblige the pre-frontal cortex to queue responses (Miller and Wallis, 2008); indeed, when people are performing two different tasks at the same time, it means that one task involves a decision-making process, while the other has become automatic or routine (Schumacher *et al.*, 2001).

Neurological research also seems to suggest that brain capacity is finite and that attention to one task diminishes as another is introduced (Just *et al.*, 2001). In a much quoted assertion, David Meyer put it this way: "if a teenager is trying to have a conversation on an e-mail chat line while doing algebra, she'll suffer a decrease in efficiency, compared to if she just thought about algebra until she was done. People may think otherwise, but it's a myth" (Wallis, 2006). The ability to multitask, as often attributed to young people, therefore seems to be more accurately expressed as fast-switching among the different brain areas (Small and Vorgan, 2008).

Neurological research has identified the portion of the brain responsible for this kind of activity-switching (Wallis, 2006), but little is known about the effects of constant switching between media, even if all of them are supported by just one technological device, such as a computer. It would seem that this rapid activity switching is not beneficial to young people's learning. It has been broadly shown that such rapid switching behaviour, when compared to carrying out tasks serially, leads to poorer learning results in students and poorer performance of tasks (Gopher, Armony and Greenspan, 2000). This is primarily due to the fact that switching requires a person to juggle her or his limited cognitive resources to accomplish the different tasks successfully. This juggling leads to greater inefficiency in performing each individual task, namely more mistakes are made, and it takes significantly longer as compared to sequential work (Ophira, Nass and Wagner, 2009).

Scholars have underlined that proposing media multitasking activities in learning experiences does not offer a straightforward interpretation: on one hand it can be an effective strategy against boredom, on the other it comes along with distractions. What is certain is that multitasking is a phenomenon that will not disappear – rather, it will become mainstream.

Limits of available evidence

The influence of technology use on reasoning capability and judgment has been shown to be relatively small while there are many studies regarding the influence of technology use on abilities related to information processing, reflective and critical thinking, creativity and, meta-cognitive skills in general – although these will be not discussed here. However, no research review has yet documented a positive effect on the basis of empirical research. And it may well be that this shows the need for a "neuroscience of children and media" intended to research the impact of digital media on children's brain development, a need that has only been expressed very recently (Anderson, 2007).

Most of the criticism with regard to assessing the impact of technology on cognitive skills is connected with the practice of taking measurements immediately after practicing, meaning that the cumulative effect of digital media is not sufficiently examined. Although the reviewed findings reflect a mostly intentional training of cognitive skills and not the effects from everyday use of digital technologies, the same mechanisms can be expected to operate in natural settings. Thus, most of the uses of digital media involve complex processes and are influenced to a large extent by structural design features, perception and cognitive properties.

However, constructing mental models and internalising concepts from using different computer applications and the Internet also influences how young people think, as recent experimental research indicates. The results of a number of experimental studies (Sparrow, Liu and Wegner, 2011) suggest that when faced with difficult questions, people are primed to think about computers and that when people expect to have future access to information, they have lower rates of recall of the information itself and enhanced recall instead for where to access it. The Internet has become a primary form of external or transactive memory, where information is stored collectively outside ourselves.

Social values and attitudes

One of the major questions regarding the role of digital technologies in young people's lives concerns how socialisation processes and social behaviour are influenced by the increasing spread and use of computers and the Internet. The availability of technology and some criticised features of digital content, particularly in video games, such as the stereotyping of women and minorities and the enforcement of violence, have raised concerns about the long-term effects on the identity and social development of young people. However, the question of the actual impact of using digital media on young people's skills for building and maintaining social relationships, and on their views, attitudes and behaviour, is also a question of the differences in media availability and use among social groups.

Belonging to a social group with specific values, lifestyle, cultural practices and preferences naturally affects media socialisation, and while for young children the family circle is a deciding factor, youth culture gains importance for adolescents. The social milieu approach in examining the use of media by young people has already proven useful for explaining television habits and preferences; however, it has not been explored with regard to digital technologies, and research in this direction can be expected to contribute to the development of this field. The different uses of computers and the Internet according to age, gender and educational level lead to additional inequalities. Thus, digital media use is determined by age, gender and educational characteristics as well as environmental influences such as the values and preferences of parents and peers. Studies on "media ecologies" provide evidence that media usage, media culture, and media language affect the way in which children and adolescents live and perceive their self-representation with respect to the social projection: new styles of participation are emerging and the concept itself of group of peers changes, being – potentially – ever connected thanks to social networking, where is the limit between getting together and being together (see also Ito *et al.*, 2009)?

The various possible uses deepen the discrepancies between social groups through the respective development of competencies, learning styles and strategies, as well as attitudes and values, which then affect outcomes, lives and career paths. So, in many different ways, it appears to be the case that it is not so much that technology use has an influence on attitudes and social values, but rather that attitudes and social values affect technology use.

The paradox of socially-connected isolation

The increase in use of computers and the Internet by children and adolescents has also been connected to concerns isolationist tendencies and hindered development of social competencies. Indeed, in most OECD countries the proportion of children and young people with digital devices in their rooms has increased, but, at the same time, the devices are used more frequently than before for communication purposes through a variety of applications and even more so since the emergence of Web 2.0 social applications. These are especially popular among teenage girls. As a result, young people tend to project an image of isolation – they appear to be concentrating on themselves – but, in fact, most of the time they are interacting socially with others while being continuously online or texting messages.

Despite the clearly increasing digital media use for social purposes, it is not immediately obvious how this affects interpersonal skills and social resources. On the one hand, it becomes clear that social and communication applications are used primarily to keep up with close friends and close family members, and the use of the computer for e-mail in these online relationships supplements the telephone and face-to-face visits but rarely replaces older modes of communication. On the other hand, uploading video clips on YouTube or participating in online games like *World of Warcraft* might indicate that the Internet also opens up opportunities for participation in global networks, thus extending the normal boundaries of social networks.

Nevertheless, it is clear that the time allotted to family interaction has been reduced. When families adopt yet another solitary medium into their home, time spent with the computer takes away from time spent with other family members. However, in the context of most OECD families, computers and the Internet are just another addition to an already complex media environment; if, as a result, family time is displaced, the computer is not to be blamed. Therefore, computer use can potentially have a negative impact on social interaction in the family (unless parents do not care about the amount of time spent at the computer), but, as a counterweight, it may increase feeling of belonging with peers (Krcmar and Strizkakova, 2006).

The lack of digital culture and the limits of self-evaluation

Learning by doing, by playing or by trial-and-error, which is typical of ICT-mediated access to knowledge and socialisation, are likely to reduce young people's ability to honestly self-evaluate their skills in media use. As shown in a recent Australian study the NML are not are not big users of Web 2.0 technologies (Kennedy *et al.*, 2007). A possible explanation is that "even though students are usually interested in new activities and new 'fashions', they tend to do always the same thing on Internet, they follow a pattern" (Reia *et al.*, 2006). This can be explained by a lack of devoted training on media culture and an insufficient and inadequate appropriation of the net-life (Bevort and Breda, 2008) due to the gap of Internet use at home (very often without their parents' assistance) versus use at school.

The outcome is an elementary use of ICT, namely a lack of a set of proper media competencies, which could come from usages assisted by competent adults. This isolated socialisation via digital technologies unbalances positively the objectivity of young people in self-evaluation as media users; for example "79% consider they know well how the Internet works; but only 52% of them are able to evaluate the information they retrieve".[7]

The effects of video games

Computer and video games are the fastest growing entertainment segment (DFC, 2004) worldwide, with a market worth 35 billion USD worldwide in 2008. As a whole, already in 2000 they surpassed all Hollywood revenues (RocSearch, 2005). In 1993, the average United States 14-year-old spent a maximum of 6 hours per week playing video games, but in fact 12% of boys and 37% of girls did not play video games at all (Funk, 1993). A decade later, in 2003, the average United States 14-year-old spent on average 17 hours per week on video games, and almost 10% spent more than 30 hours per week (Sherry *et al.*, 2003).[8]

Like any medium, video games are a channel of communication whose effects vary widely with the content of the specific game in question. Experimental, longitudinal and meta-analytic data indicate that playing violent video games increases aggression, hostility and aggressive thoughts (C.A. Anderson, Gentile and Buckley, 2007). Because research consistently shows that most popular video games are violent (Dill *et al.*, 2005; Lachlan, Smith and Tamborini, 2005), and because of the potential harm to children, youth and society of this negative influence, much video game research has focused on the effects of violent video games. Modelled after the extensive literature on television violence effects, the literature on the effects of violent video games shows a consistent link between violent video game play and aggression (C.A. Anderson *et al.*, 2003).[9]

The debate about the effects of playing computer games with violent content has been heated again by recent findings, which tend to attract enormous media attention when they point to the negative effects of playing video games. Spitzer (2007) stated that violent content influences brain processes and is related to the development of aggressive behaviour which would eventually lead to an increase in criminal events in future. Such findings replicate what an early meta-analysis on the research on violent video games (Anderson and Bushman, 2001) found: exposure to violent video games is positively associated with increased levels of aggression. And, in fact, if only those studies with the soundest methodological approaches were used, results showed even stronger correlation (C.A. Anderson, 2004). Moreover, effects sizes seem to have increased over time, with more current studies showing stronger effects, presumably due to the greater realism of today's games.

On the other hand, researchers have recently begun to focus attention on stereotypical portrayals of women and minorities in video games and the adverse effects of these characterisations (Brenick *et al.*, 2007). It is now clear that there is a positive association between violent video game play and misogynist attitudes including attitudes supporting violence against women.[10] A number of studies found that women are under-represented, stereotyped and objectified in video games. Some of this research indicates that minorities are also stereotyped in video games.[11]

Video games convey interpretations of reality, social relations and events, but often in simplified and stereotyped form. Several theories explain how exposure to video games, especially violent games, can lead to imitative behaviour. The bestselling video games do have a negative impact on social attitudes towards violence and sexual harassment by lowering the threshold of tolerance. However, it appears to be widely assumed that only on very rare occasions are players under the illusion that their behaviours in the game world are transferable to real world. At least, this seems to be the view shared by most parents, educators and, possibly, policy makers, who do not consider it a matter of serious concern that the video game genre most preferred by children up to the age of 18 is action or combat (42%) (Rideout, Foehr and Roberts, 2010), probably because they have not received a clear warning message from the research community.

Unfortunately, there is scarce evidence regarding the possible positive effects of video games, probably because researchers are not interested in them, and media may not pay as much attention to such findings. However, existing evidence indicates that moderate game-playing does not significantly impact children's social skills either positively or negatively as studies often found no differences in the sociability and social interactions of computer games players versus non-players (Subrahmanyam *et al.*, 2000). So, the issues at stake are: what is considered to be moderate use and what kind of adult supervision do video games require?

Finally, the importance of parental involvement has to be underlined once again. Statistics on young Americans demonstrated that the amount of time spent on media use is effectively reduced when parents build the family media environment with awareness and decide clearly the rules of usages as seen in the Figure 4.1. It is evident that a control on timing implies a reduction of likelihood of side effects of media use (Rideout *et al.*, 2010).

Finally, there is an increasing body of research, as well as public support, for new categories of video games intended to use the same technologies but with a clear educational purpose. The so-called serious games are based on concepts such as flow, immersion or user-led activity that can be extremely helpful when designing learning environments (Ritterfeld, Cody and Vorderer, 2009). And there is promising research intended to

Figure 4.1. **The impact of media rules**

Amount of total media exposure in a typical day among children who say they have:

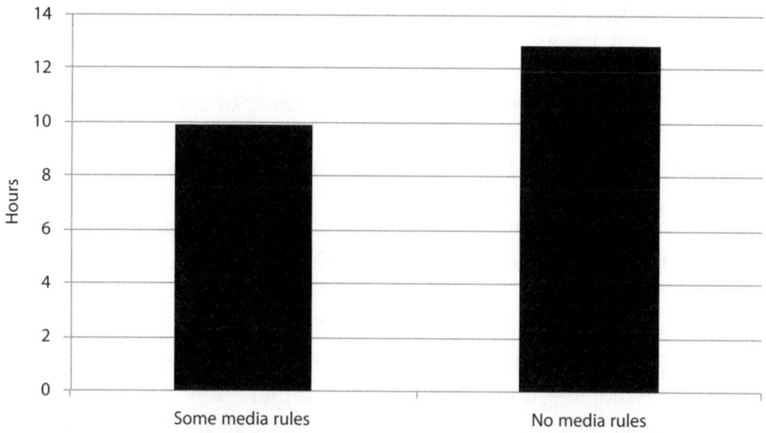

Source: A Kaiser Family Foundation study (2010), Generation M2: Media in the Lives of 8- to 18-year-olds report. *www.kff.org/entmedia/upload/8010.pdf.*

assess the impact of video games on educational outcomes (Mayo, 2009; Ritterfeld *et al.*, 2009).Yet, although there is a lot of potential in them, there are some important limitations as well, imposed by the fact that the investments required for video games development are not easily available in the educational market – not to mention the reluctance that video games as educational tools may raise in teachers.

Although much exposure to digital media by children and young people is typically intended for entertainment purposes only, such an exposure may have exhibited effects far beyond simple amusement. An increasing amount of research indicates that playing action video games is associated with a number of enhancements in vision, attention, cognition, and motor control (Green and Bavelier, 2008). A few examples of research evidence in this direction include the following: action video game experience heightens the ability to view small details in cluttered scenes and to perceive dim signals (Li *et al.*, 2009); keen players display enhanced top-down control of attention and choose among different options more rapidly (Dye, Green and Bavelier, 2009); frequent players also exhibit better visual short-term memory (Boot *et al.*, 2008), and can more flexibility switch from one task to another (Colzato *et al.*., 2010; Karle, Watter and Shedden, 2010). Needless to say, these research findings have been found to have real-world applications particularly in the area of professional training but not so often in school education.

Overall conclusion

Although the existing evidence does not offer conclusive answers to questions regarding the effects of different uses of technology on the social development and behaviour of young people, it indeed indicates that its potential harmful impact cannot be overlooked. Therefore, it is necessary to explore how these negative effects can be lessened or prevented. Despite the attempts of governments to limit the possible access of children to violent content, this can easily be circumvented with the help of siblings, friends and even parents. It seems that control of children's and adolescents' use of digital media by parents is of primary importance; thus, parents need to be alerted to the dangers of the opportunities new technologies create. This could also be a subject of media education directed towards parents.

However, the most beneficial option would be to equip young people with the knowledge and competencies that would enable them to navigate in online spaces and virtual worlds with less harm rather than relying on external control. Such interventions are particularly suitable for schools because of the potential they offer to reach every child or adolescent. Clarifying the functions of digital media and the differences between conflict resolution strategies and social actions in virtual worlds and those in real life should be an integral part of training programmes. The identification of risk groups and effective interventions are the first steps toward more practice-oriented research.

Notes

1. Because of its nature and purpose, the NML project does not focus on the eventual effects of technology on physical well-being, where research has already highlighted that children's extended computer use may be linked to an increased risk of obesity, seizures and hand injuries (Subrahmanyam *et al.*, 2000).

2. The contents of this section are mostly, but not only, based on the research reviews conducted in a number of OECD countries in clusters according to language or cultural affinity (English, French, German, and Spanish languages, plus a review of the Scandinavian countries). Korea contributed also with one national review.

3. The Flynn effect is the rise of average Intelligence Quotient (IQ) test scores, an effect seen in most industrialised countries, although at greatly varying rates.

It is named after James R. Flynn, who did much to document it and promote awareness of its implications. This increase has been continuous and roughly linear from the earliest days of testing to the present.

4. Flynn gives as an example the following question: "What do a dog and a rabbit have in common?" A modern respondent might say they are both mammals (an abstract answer), whereas someone a century ago might say that you catch rabbits with dogs (a concrete answer).

5. As the authors openly state, "our search criteria focused on the literature examining the efficacy and effectives, rather than the negative outcomes of computer applications. As a result, the results were generally positive in nature because the 'negative' findings were limited" (p. 32).

6. The authors said that "there is always a possibility that advocacy plays a role in the aggregation and reporting of data" (p. 29).

7. From the international research Mediappro, involving about 9 000 young people aged between 12 and 18 (7 400 in Europe, plus 1 350 in Quebec, Canada). See: *http://edu.of.ru/attach/17/39925.pdf#page=89* for the final report.

8. Gender differences are striking in this respect. Among children ages 2 to 7, boys are 25% more likely than girls to lay video games on a regular basis, whereas male teenagers are 49% more likely to play than their female counterparts (Rideout, Vandewater and Wartella, 2003).

9. For example, Carnagey and Anderson (2005) found that when a car racing game rewarded players for violent acts, those players were more likely to attack an opponent than when the same game punished players for aggression. Konijn, Bijvank and Bushman (2007) found that adolescent boys who identified with aggressive characters in immersive, realistic games were most aggressive, going so far as to blast opponents with noise levels they believed would cause permanent hearing damage.

10. For example, Dill *et al.* (2005) found that youth exposed to sexist images of video game characters were more likely to accept rape myths (such as the idea that women enjoy sexual force, that men should dominate women sexually and that women who say "no" are simply engaging in "token refusals") than youth exposed to images of professional men and women.

11. For example, Dill *et al.* (2005) found that Middle Easterners were over-represented as targets of violence in video games. Burgess *et al.*, in press, found that male African-American video game characters are stereotyped as athletes and "gangstas" or "thugs" who are more likely to use guns – particularly extreme guns – than characters of other races. Furthermore, most Asian women are represented as non-aggressive beauties and most Asian men (fully 75%) are shown using martial arts.

References

Ackerman, P.L., M.E. Beier and K.R. Bowen (2002), "What we really know about our abilities and our knowledge", *Personality and Individual Differences*, Vol. 33, pp. 587-605.

Anderson, C.A. (2004), "An update on the effects of playing violent video games", *Journal of Adolescence*, Vol. 27(1), pp. 113-122.

Anderson, C.A. (2007), "A neuroscience of children and media?" *Journal of Children and Media*, Vol. 1(1), pp. 77-85.

Anderson, C.A. *et al.* (2003), "The influence of media violence on youth", *Psychological Science in the Public Interest*, Vol. 4(3), pp. 81-110.

Anderson, C.A. and B.J. Bushman (2001), "Effects of violent video games on aggressive behavior, aggressive cognition, aggressive affect, physiological arousal, and prosocial behavior: A metaanalytic review of the scientific literature", *Psychological Science in the Public Interest*, Vol.12, pp. 353-359.

Anderson, C.A., D.A. Gentile and K.E. Buckley (2007), *Violent Video Game Effects on Children and Adolescents, Theory, Research, and Public Policy*, Oxford University Press, New York.

Anderson, J.R. (2000), *Cognitive psychology and its implications*, Freeman Publishing Company, New York.

Arnseth, H.C. and S.R. Ludvigsen (2006), "Approaching institutional context: Systemic versus dialogic research in CSCL", *International Journal of Computer-Supported Collaborative Learning* (1), pp. 167-185.

Atkinson, N.L. *et al.* (2001), *Technology and Child Development, Part I: a Ten-Year Review of Reviews*, The Center for Child Well-being, Decatur, GA.

Baddeley, A.D. and R.H. Logie (1999), "Working memory: The multiple component model" in A. Miyake and P. Shah (eds.), *Models of working memory: Mechanisms of active maintenance and executive control* (pp. 28-61), Cambridge University Press, New York.

Bavelier, D., C.S. Green and M.W.G. Dye (2010). "Children, Wired: For Better and for Worse", *Neuron,* Vol. 67, Issue 5, pp. 692-701, *www.cell.com/neuron/abstract/S0896-6273(10)00678-1?switch=standard.*

Bevort, E. and I. Breda (2001), "Les jeunes et Internet, Représentations, utilisations, appropriation", *www.clemi.org/jeunes_internet.html.*

Boot, W.R. *et al.* (2008), "The effects of video game playing on attention, memory, and executive control", *Acta Psychologica, Vol.* 129, Issue 3, pp. 387-398, *http://dx.doi.org/10.1016/j.actpsy.2008.09.005.*

Bracken, C.C. and M. Lombard (2004), "Social presence and children: Praise, intrinsic motivation, and learning with computers", *Journal of Communication* (54), pp. 22-37.

Brenick, A. *et al.* (2007), "Social reasoning about stereotypic images in video games: Unfair, legitimate, or 'just entertainment'?", *Youth and Society* (38), pp. 395-419.

Burgess, M. *et al.* (2011), "Playing with Prejudice: The Prevalence and Consequences of Racial Stereotypes in Videogames", *Media Psychology*, Vol. 14, pp. 289-311.

Carnagey, N.L. and C.A. Anderson (2005), "The effects of reward and punishment in violent video games on aggressive affect, cognition and behavior", *Psychological Science in the Public Interest,* Vol.16, pp. 882-889.

Colzato, L.S. *et al.* (2010), "DOOM'd to switch: superior cognitive flexibility in players of first person shooter games", *Frontiers in Psychology,* Vol. 1, pp. 1-5, *http://doi: 10.3389/fpsyg.2010.00008.*

DFC Intelligence (2004), "The Business of Computer and Video Games 2004", *DFC Intelligence*, San Diego, CA.

Dill, K.E. *et al.* (2005), "Violence, sex, race and age in popular video games: A content analysis", in E. Cole and J.H. Daniel (eds.), *Featuring females: Feminist analyses of the media*, American Psychological Association, Washington, DC.

Dux, P. E., *et al.* (2006), "Isolation of a central bottleneck of information processing with time-resolved fMRI", *Neuron*, Vol. 52(6), pp. 1109-1120.

Dye, M.W., C.S. Green and D. Bavelier (2009), "Increasing Speed of Processing With Action Video Games", *Current Directions in Psychological Science,* Vol. 18, pp. 321–326, *http://vision.psych.umn.edu/users/csgreen/Publications/dye_CDiPS09.pdf.*

Eimer, M., J. Van Velzen and J. Driver (2002), "Crossmodal interactions between audition, touch and vision in endogenous spatial attention: ERP evidence on preparatory states and sensory modulations", *Journal of Cognitive Neuroscience*, Vol. 14, pp. 254-271.

Flynn, J.R. (2007), *What is Intelligence?: Beyond the Flynn Effect*, Cambridge University Press, Cambridge.

Funk, J. B. (1993), "Reevaluating the impact of videogames", *Clinical Pediatrics,* Vol. 32, pp. 86-90.

Gopher, D., L. Armony and Y. Greenspan (2000), "Switching tasks and attention policies", *Journal of Experimental Psychology: General*, *129*, pp. 308-229.

Green, C.S. and D. Bavelier (2008), "Exercising your brain: a review of human brain plasticity and training-induced learning", *Psychology and Aging*, Vol. 23, no. 4, pp. 692-701, *www.ncbi.nlm.nih.gov/pubmed/19140641.*

Greenfield, P.M. (2009), "Technology and informal education: what is taught, what is learned", *Science*, Vol. 323(5910), pp. 69-71

Greenfield, P.M. (1998), "The cultural evolution of IQ", in U. Neisser (ed.), *The raising curve: Long-term gains in IQ and related measures*, American Psychological Association, Washington, DC.

Ito, M., *et al.* (2009), *Hanging out, messing around, and geeking out: Kids living and learning with new media*, MIT Press, Cambridge, MA.

Just, M.A. *et al.* (2001), "Interdependence of Nonoverlapping Cortical systems in Dual Cognitive Tasks", *Neuroimage*, Vol. 14(2), pp. 417-426.

Kaiser Family Foundation (2010), *Generation M2: Media in the Lives of 8-Year-Olds, www.kff.org/entmedia/upload/8010.pdf.*

Karle, J.W., S. Watter and J.M. Shedden (2010), "Task switching in video game players: Benefits of selective attention but not resistance to proactive interference", *Acta Psychologica,* Vol. 134, pp. 70-78, *www.ncbi.nlm.nih.gov/pubmed/20064634.*

Kennedy, G., *et al.* (2007), "The net generation are not big users of Web 2.0 technologies: Preliminary findings", paper presented at the ASCILITE conference, Singapore.

Klingberg, T. (2009), *The overflowing brain: information overload and the limits of working memory*, Oxford University Press, New York.

Koeppl, P.T. *et al.* (2001), *Technology and Child Development, Part II: Lessons from Empirical Research*, The Center for Child Well-being, Decatur, GA.

Konijn, E.A., M.N. Bijvank and B.J. Bushman (2007), "I wish I were a warrior: The role of wishful identification in effects of violent video games on aggression in adolescent boys", *Developmental Psychology*, Vol. 43, 1038-1044.

Krcmar, M. and Y. Strizkakova (2006), "Computer-Mediated Technology and Children", in C.A. Lin and D.J. Atkin (eds.), *Communication Technology and Social Change: Theory and Implications,* Routledge, New York.

Lachlan, K.A., S.L. Smith and R. Tamborini (2005), "Models for aggressive behavior: The attributes of violent characters in popular video games", *Communication Studies* Vol. 56, pp. 313-329.

Li, R. *et al.* (2009), "Enhancing the contrast sensitivity function through action video game training", *Nature Neuroscience* Vol. 12, pp. 549-551, *www.nature.com/neuro/journal/v12/n5/abs/nn.2296.html.*

Mayo, M.J. (2009), "Video Games: A Route to Large-Scale STEM Education?", *Science,* Vol. 323(5910), pp. 79-82.

Miller, E.K., Wallis, J.D. (forthcoming), "The prefrontal cortex and executive brain functions", *Fundamental Neuroscience,* Vol. 4.

Mößle, T. *et al.* (2006), "Mediennutzung, Schulerfolg, Jugendgewalt und die Krise der Jungen", *Zeitschrift für Jugendkriminalrecht und Jugendhilfe,* Vol. 3, pp. 295-309.

Neisser, U. (1998), *The raising curve: Long-term gains in IQ and related measures*, American Psychological Association, Washington, DC.

Newcombe, N.S. and J. Huttenlocher (2000), *Making Space: The Development of Spatial Representation and Reasoning*, MIT Press, Boston, MA.

Niederhauser, D.S. *et al.* (2000), "The influence of cognitive load on learning from hypertext", *Journal of Educational Computing Research*, Vol. 23(3), pp. 237-255.

Ophira, E., C. Nass and A.D. Wagner (2009), "Cognitive control in media multitaskers", *Proceedings of the national academy of sciences of the United States of America*, Vol. 106(33).

Reia, *et al.* (2006), "The appropriation of new media by youth", Mediappro report, *www.mediappro.org/publications/finalreport.pdf.*

Rideout, J., *et al.* (2010) *Generation M2 : media in the lives of 8- to 18-year-olds,* Henry J. Kaiser Family Foundation, Menlo Park, California, USA.

Rideout, V.J., D.F. Roberts and U.G. Foehr (2005), *Generation M: Media in the lives of 8-18 year-olds*, The Henry J. Kaiser Family Foundation, Menlo Park.

Rideout, V.J., U.G. Foehr, and D.F. Roberts (2010), *Generation M2: Media in the Lives of 8- to 18-Year Olds*, Henry J. Kaiser Family Foundation, Menlo Park, CA, *www.kff.org/entmedia/upload/Generation-M-Media-in-the-Lives-of-8-18-Year-olds-Report.pdf.*

Ritterfeld, U. *et al.* (2009), "Multimodality and Interactivity: Connecting Properties of Serious Games with Educational Outcomes", *CyberPsychology and Behavior*, Vol. 12(6), pp. 691-697.

Ritterfeld, U., M. Cody and P. Vorderer (eds.) (2009), *Serious Games. Mechanisms and Effects*, Routledge, London.

RocSearch (2005), *Videogame Industry*, RocSearch, London.

Schumacher, E. H., *et al.* (2001), "Virtually perfect time sharing in dual-task performance: Uncorking the central cognitive bottleneck", *Psychological Science*, Vol. 121, 101-108.

Sherry, J. L. *et al.* (2003), "Why do adolescents play videogames?", Developmental stages predicts videogame uses and gratifications, game preference and amount of time spent in play, paper presented at the Annual Convention of the International Communication Association.

Sherry, J.L. *et al.* (2003), *Why do adolescents play videogames? Developmental stages predicts videogame uses and gratifications, game preference, and amount of time spent in play*, paper presented at the Annual Convention of the International Communication Association.

Shimojo, S. and L. Shams (2001), "Sensory modalities are not separate modalities: plasticity and interactions", *Current Opinion in Neurobiology*, Vol. 11(4), pp. 505-509.

Small, G.W. and C. Vorgan (2008), *IBrain : surviving the technological alteration of the modern mind*, Collins Living, New York.

Souvignier, E. (2001), "Training räumlicher Fähigkeiten", in K. Klauer (ed.), *Handbuch Kognitives Training* (pp. 293-319), Hogrefe, Göttingen.

Sparrow, B., J. Liu and D.M. Wegner (2011), "Google Effects on Memory: Cognitive Consequences of Having Information at Our Fingertips", *Science*, Vol. 333(6040), pp. 15-20.

Spitzer, M. (2007), *Vorsicht Bildschirm! Elektronische Medien, Gehirnentwicklung, Gesundheit und Gesellschaft*, dtv, München.

Subrahmanyam, K. *et al.* (2000), "The impact of home computer use on children's activities and development", *Children and Computer Technology* (10), pp. 123-144.

Wallis, C. (2006), "The Multitasking Generation: they're emailing, IMing and downloading while writing the history essay. What is all that digital juggling doing to kids' brains and their family life?", *Time* (167), pp. 48-55.

Wright, P.C., R.E. Fields and M.D. Harrison (2000), "Analyzing human-computer interaction as distributed cognition: The resources model", *Human-Computer Interaction*, Vol. 15(1), 1-41.

Chapter 5

Are learners' expectations changing?

The argument of dramatic changes in students' learning expectations as a result of their being New Millennium Learners is quite often mentioned as one of the most powerful drivers for change in education. However, this proposition can hardly be backed with evidence. Contrary to what many voices have suggested, students cannot be said to have dramatically changed their expectations about teaching, learning and technology: although they value the convenience and the benefits that they get with technology, their preferences are still for traditional face-to-face teaching where technology improves current practices and results in higher engagement, a more efficient resolution of learning tasks and increased outcomes. If those gains do not become apparent to students, then reluctance emerges. The reasons for such reluctance might be related to the uncertainty, disruptiveness and discomfort that discrete technology-based innovations not clearly leading to learning improvements may cause to them. Therefore, the idea that students would be the strongest supporters of radical transformations in education, as attractive as it may seem, is not yet supported by research evidence. But this might be changing as the more rewarding experiences students get, the more likely they are to become supportive of technology-based innovations. A clear implication is that teachers will have to lead the way.

It is commonly stated that students' attitudes and expectations regarding learning have evolved radically from previous generations. Some twenty years ago, when computers began to be seen as a tool that could potentially improve the quality of teaching, school teachers began to experiment with the idea of supplementing traditional teaching and learning activities with educational applications and digital resources which later on, thanks to the Internet, became increasingly available. Up to then, teachers could be said to have been in command of the opportunities for technology-mediated educational innovations, and, for most students, schools were the only place where they could get access to a computer. But when both computers and the Internet started to enter pupils' homes and increasingly became a standard commodity for most families, students' technology-related competencies grew exponentially, even by self-learning, as the PISA survey revealed in 2003 (OECD, 2005), largely outperforming those of their teachers.

A rationale for evolving expectations

The assumption that students' expectations have changed dramatically and are completely different from the ones held by their teachers seems to be already considered a given fact in current debates about technology and education. Such dramatic changes would have an impact particularly on the following aspects:

- range of technology devices and services available to them;
- frequency of their use;
- range of possible teaching and learning activities;
- opportunities for collaborative work and networking;
- communication skills involved (including a reinterpretation of written language);
- degree of learning personalisation; and
- standards of digital quality, in terms of interactivity and use of multimedia resources.

A study from Pew Internet and American Life found that in the United States more than half of the 12 million teens online create original material for the web, whether it's through a blog or a home page, with original artwork, photos or video (Lenhart *et al.*, 2007). This translates into a relevant proportion of higher education students contributing to blogs, photo or video sites, thus becoming content producers. This in turn may have an effect on their expectations – for instance, most British prospective university students (79%) would expect to take their own computer to university with them and to be able to use it by logging on the university network (81%) (Ipsos Mori, 2007).

Teachers' estimates of students' expectations

It could be reasonably expected that today's students are more willing to use technology in learning activities than education institutions allow them to do. To what extent this contrasting situation makes them feel disappointed with teaching practices, or even increasingly disaffected from schooling, has not been thoroughly investigated yet, but some indications exist pointing to a growing gap between students' and teachers' perceptions regarding the quality of the educational experience (BellSouth Foundation, 2003) and, moreover, how students conceptualise their new millennium learning styles and expectations and teachers' beliefs about these.

However, it turns out that teachers tend to overestimate the extent to which their students have the type of expectations characteristic of New Millennium Learners. This may be an effect of the establishment of a dominant view, which infers from students' current levels of digital media exposure and connectedness, a totally different approach to learning and radically new expectations when it comes to teaching, learning and the role that technology has to play in them. This is not the case – at least not yet.

A clear indication of the disparity between views is reflected in Figure 5.1 covering six European countries (France, Germany, Italy, the Netherlands, Spain and Sweden) based on a European comparative study (Lam and Ritzen, 2008).[1] The figure presents the percentage of disagreement between how university students see themselves and how teachers think of students, in a range of learning characteristics which are assumed to be attached to a new millennium learning style.

Teachers appear to slightly overestimate what students claim to be most of the learning characteristics usually attached to the concept of New Millennium Learners. The differences are high with respect to the willingness to benefit from learning by doing, and the preference for visual learning (20% and 15%, respectively). On the other side, teachers tend to underestimate only students' preferences for social and interactive learning as well as for carrying out several activities simultaneously – but the differences are much less significant. There are also important differences between countries, as Figure 5.2 indicates.

It is in the Southern European countries covered in this study (France, Italy and Spain) where the disparities between students' self-perceptions and teachers' views of them are the highest. On the whole, the disparities in Germany, the Netherlands and Sweden are almost irrelevant. To sum up, the comparison between students' self-perceptions and teachers' views about

Figure 5.1. **Percentage of disparity between university students' self-perceptions regarding their own learning characteristics and teachers' views**

Average values in six European countries (France, Germany, Italy, the Netherlands, Spain and Sweden), 2008

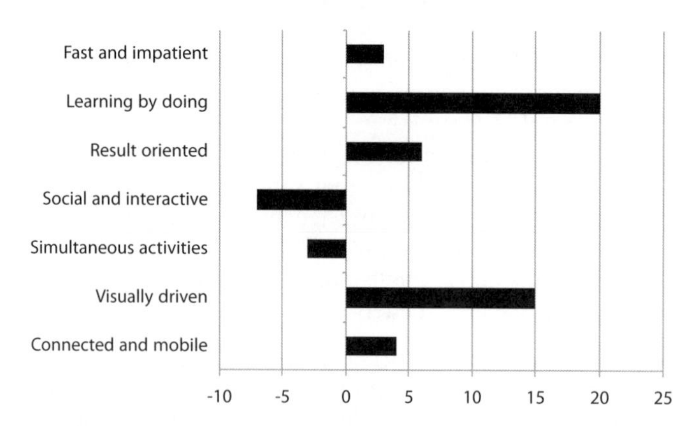

Source: Lam and Ritzen, 2008. Data are based on questionnaires administered face-to-face. The percentage expresses the relative difference between teachers' and students' perceptions. Positive values indicate overestimates by teachers, while negatives indicate underestimates by teachers, in comparison to students' declared self-perceptions.

Figure 5.2. **Percentage of disparity between students' self-perceptions and teachers' views across countries**

Average values per country, 2008

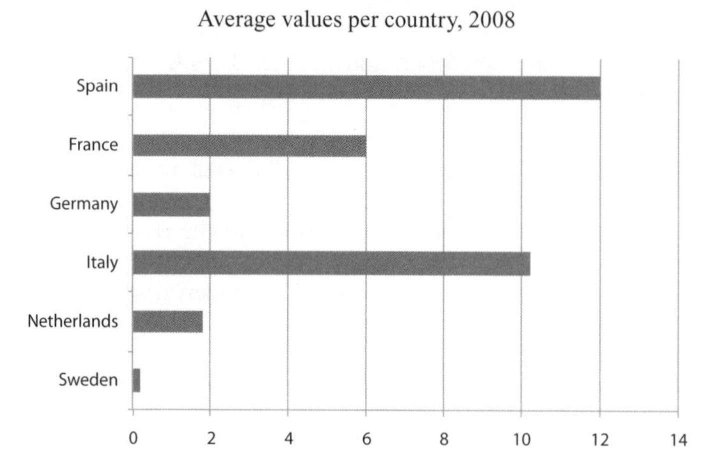

Source: Lam and Ritzen, 2008. Data are based on questionnaires administered in interviews. The percentage expresses the relative difference between teachers' and students' perceptions. Positive values indicate overestimates by teachers, while negatives indicate underestimates by teachers, in comparison to students' declared self-perceptions.

students' learning preferences and styles seems to suggest a clear degree of teacher awareness of those preferences, and even a certain tendency to overestimate some of them. Therefore, teacher awareness appears to be already in place, although it differs depending on country.

Are students more reluctant than expected to adopt technology in teaching?

On the whole there is little empirical evidence regarding the so often assumed shift in students' demands and expectations. Although student surveys have been in place for a long time in a number of OECD countries, including Australia, France, Norway, Sweden, the United Kingdom and the United States,[2] the issues related to expectations regarding technology in teaching have not been taken into account except in surveys where the main topic is precisely technology adoption.[3] International comparative evidence is even scarcer, and sometimes the nature of the methodology used does not allow room for generalisations. However, there are a few studies, with very limited and unrepresentative samples, which might be taken as indicative of what may be occurring.

What emerges from available data is that students appear far more reluctant towards technology adoption in education than their levels of digital media exposure and connectedness would suggest. In general, they welcome uses and applications that are intended to provide more convenience (for example, access to course guidelines, notes or background documents) or that improve their productivity in academic tasks (for example, online databases or virtual libraries). Other than this, they advocate use of technology in teaching that supplements rather than changes the traditional models, and they certainly show a clear preference for face-to-face teacher or tutor relationship over computer-mediated communications.

Students set limits to technology use in the classroom

To begin with, the main reasons students may be keen for the use of technology in their courses are not so much related to their willingness to see teaching and learning radically transformed as to the added value of convenience. This was already pointed out by Caruso and Kvavik (2006), who found that the most valuable reason for using technology in courses is precisely convenience (51% of students), followed by the ability to more easily manage course activities (19%), and, to a much lesser extent, to provide opportunities to enhance learning (15%) and to communicate with peers and teachers (11%). Accordingly, from the student perspective, technology is useful because of the convenience and control it provides rather than for its transformative power.

Overall, European students clearly appear to desire more use of technology in their courses although a significant number, roughly one-fifth, remain unsure (Flather and Huggins, 2004). This may suggest some concern that the benefits of improved communication may also lead to less direct contact with staff, or that remote or distance learning may replace some traditional teaching methods. Further, these research results may reflect student concern that the way in which technology is actually being used by teachers in fact requires students to do even more things or may be asking them to participate in unexpected activities or activities the purpose of which is not well understood and the added value of which is not clear to them – or not well explained by instructors. Another European study (Spot+, 2002) found that although university students held a fairly positive view of the different advantages that ICT can bring to education, they also had a similarly positive attitude towards learning via traditional education methods and one which questioned the value of ICT in education. A closer inspection of the answers on the individual questions reveals that university students were especially interested in the use of ICT for purposes of information exchange, such as "to ask questions of experts and relevant people no matter where they are" and "to share information and ideas with people who have similar interests."

With respect to explicit learning purposes, the students expressed a stronger preference for traditional education methods (defined as printed text and a classroom setting) than for ICT-based methods. In many ways it may well be that student expectations regarding technology adoption in teaching are less supportive of innovations than is commonly taken for granted.

There are clear indications that students' main assumption is that teaching is about conveying knowledge from the teacher to the learner, from a position of authority. If ICT is to be used in an educational context, students specifically expressed doubts about the quality of the human interaction when there is no face-to-face contact. Moreover, 21% of the respondents indicated that they did not know whether "small-group learning may become disorganised in online courses", 14% did not know whether "learning with ICT is very time-consuming", and 13% did not know whether "ICT can improve their learning" Higher Education Academy (2007). A recent survey of prospective students in the United Kingdom found that four-fifths (80%) felt that the quality of teaching at university was more important than the IT provision (Ipsos Mori, 2007).

This is seen across the board – high or low ICT use does not necessarily correlate with perceived importance of quality of teaching over ICT provision. ICT is seen as a supplement to teaching, not as a substitute for the personal interaction to which they are accustomed. This might indicate that, due to the lack of experience with ICT, students expressed their opinion of its use in education rather cautiously, leading them to state a higher preference for traditional education methods, which are familiar to all students. This means that younger students have a stronger preference for traditional education methods and a more negative attitude towards using ICT than students who are a few years older and have already passed some years at university or in work, at least those in those age ranges for which there is evidence, *i.e.* 15-24.

Most students seem to see technology improving their learning through giving them more access to data and research resources, rather than by imagining totally new methods of teaching, learning or interacting with peers and teachers. This mirrors their understanding of how ICT works at school and home – and it also mirrors the experience they have had so far at school, a traditional teacher and pupil environment. They find it hard to imagine other kinds of interactions and engagements. So, when British students were asked, they indicated their preference for the traditional teacher/pupil environment. As the report concludes, "the face-to-face teaching quality was felt to be the most visible sign of the university's value for money – it's what they believe they are paying for" (Ipsos Mori, 2007, p. 25). In fact, it may well be concluded that students in the United Kingdom are convinced of the benefits of technology adoption in universities, provided that it is used to support established methods of teaching and administration, and not to change them dramatically; to act as an additional resource for research and communication and to be a core part of social engagement and facilitate face-to-face friendships at university. A companion study done one year later, also in the United Kingdom, with first-year students found that face-to-face interaction is still seen as the best form of teaching, fitting well with the prevalent student view about what teaching should be. Therefore they may feel uncomfortable when teachers try to relate to them in a flat, non-hierarchical structure (such as getting involved with personal Facebook accounts). The use of ICT in teaching is now perceived to be a good thing but only as long as it is done well. Face-to-face interaction supported by inefficient or inept use of technology is worse than using none (Ipsos Mori, 2008).

All this is congruent with observations made, for instance by Oblinger and Hawkins (2005), who argued that "the assumption that students want more technology may not be valid: especially younger students are less satisfied with complete online learning than older students. The reason appears to tie to their expectation of being in a face-to-face, social environment". In a similar vein, Zemsky and Massy (2004) stated that "students do want to connect, but principally to one another; they want to be entertained, principally by games, music and movies; and they want to present themselves and their work. E-learning at its best is seen as a convenience and at its worst as a distraction – what one student called the fairy tale of e-Learning". As a recent British report has signalled, "students do not fully understand how ICT and learning can work together. They imagine and like the idea of the traditional,

Socratic, or "chalk and talk" methods with face-to-face learning" (Ipsos Mori, 2007, p. 31). Therefore, it is not surprising that European students also appear divided over the contribution that the increased usage of technology may make to the critical and intellectual abilities of students. Less than one out of ten (8%) respondents strongly agreed that ICT encouraged independent learning, whilst 9% also disagree with this statement.

ICT familiarity does not translate easily into better learning

The assumption that because students are so attached to technology in their everyday lives this warrants their full endorsement of technology in teaching and learning should be contested. The least that can be said is that it is unclear that students want their everyday technologies to be adopted in full as learning technologies.

Moreover, as Kennedy *et al.* (2008, p. 4) have pointed out, "it is not clear that emerging technologies and students' everyday skills with them will easily translate into beneficial technology-based learning". In other words, the fact that they are digitally literate does not necessarily imply that they are capable of employing technology strategically to optimise learning experiences and outcomes. As it can be read in the preface by Katz to the ECAR 2005 study (Caruso and Kvavik, 2005, p. 7), "freshman students arrive at our institutions with a set of electronic core skills. Such skills include communications (telephone, e-mail, text-messaging, and IM), web surfing (not to be confused with research skills), word processing and video gaming… These young people can make technology work but cannot place these technologies in the service of (academic) work". In fact, teachers may be expected to help students to employ technology more strategically, but do students want it?

Students have a limited understanding of how technology may support learning

Although students' expectations are often cited as a driving force for changing teaching practice, students can hardly be expected to have a consistent frame of reference for cutting edge approaches to technology-enhanced learning against which to benchmark their current learning experiences. Students' expectations of learning methods appear to be influenced more by prior experience of learning in formal situations than by students' personal use of technology outside educational settings, for instance, for informal learning or socialising (Littlejohn, Margaryan and Vojt, 2010). Furthermore. students expect technology-enhanced learning methods to reflect conventional learning, and they may be uncomfortable with the application of social technologies in educational contexts (Carey *et al.*, 2009; Harris *et al.*, 2010). Currently, there is no empirical support for the claim that young students exhibit radically different learning styles from previous generations. Instead, many findings suggest a deficit of learning literacies and a dependency on guidance from teachers amongst students (Hargittai, 2010; Jones *et al.*, 2010; Nagler and Ebner, 2009). And the paradox is that conventional forms of teaching appear to encourage students to passively consume information that is more easily available to them thanks to connectedness.

Students' expectations of learning are influenced by their teachers' approaches to teaching

Students' attitudes to learning appear to be influenced by the teaching approaches used by teachers. This finding is not new: a qualitative relationship between a lecturer's teaching methods and students' learning approaches was established through empirical studies more than a decade ago. For instance, Trigwell, Prosser and Waterhouse (1999) demonstrated that, when teachers taught using methods focused on a knowledge transmission paradigm, students were more likely to adopt a surface approach to learning. Furthermore, Virtanen and Lindblom-Ylänne (2009) emphasised that a learning approach is not a fixed student characteristic and that the same student can adopt different approaches in different contexts and even in different situations within the same context. They urged teachers to be aware of how approaches to teaching can affect students' approaches to learning. Technology adoption is not a simple binary relationship but is a complex phenomenon.

Margaryan, Littlejohn and Vojt (2010) have shown that, far from demanding that teachers change their practice, students appear to conform to fairly traditional pedagogies, albeit with minor uses of technology tools that deliver content. In fact, there is consistent evidence, even internationally, that students emphasise that they expect to be taught in traditional ways. On this basis, the claims of a growing and uniform generation of young students with radically different expectations about how they will learn seem unwarranted.

In the end, there may not be much demand from young people for school to change as technology may well play very different roles in a student's life in and out of school. Young people may be often deeply engaged with digital technology and broadly positive about that experience. However, they also identify a shared set of circumstances associated with its use in school that could be variously stressful, frustrating, threatening or devious. Clearly, the social and cognitive practices young people cultivate through the recreational use of digital devices, applications and services should not be regarded as general competencies. Rather those practices will be shaped and constrained by the particular socio-cultural settings in which such tools are used. If neither students nor teachers are firmly embracing digital technology in education, then this may reflect tensions that exist between different ambitions and expectations associated with in-school versus out-of-school cultures. If technology adoption in education remains modest, the present perspective goes beyond merely denying a rhetoric of the "digital native". Moreover, this suggests a sophisticated awareness among young people of institutional, social and moral tensions associated with modern web-based services, as well as a greater level of ambiguity of attitude than is normally recognised (Bennett and Maton, 2010; Crook, 2011).

What works for students then?

Drawing on the available evidence, there is an easy answer to the question of what works according to students. This answer refers basically to three areas: engagement, convenience and productivity. In all three areas students show a rational approach to technology in learning, possibly much more conservative than some would expect but certainly revealing concerns and issues that are not so different from those of adults.

Engagement

Students are engaged when they devote substantial time and effort to a task, when they care about the quality of their work and when they commit themselves because the work seems to have significance beyond its personal instrumental value (Newmann, 1986). That the use of ICT in education can result in higher levels of student engagement has been repeatedly demonstrated during the past thirty years (Chung and Storm, 2010; Kearsley and Shneiderman, 1998).

Technology use contributes to student engagement in particular if the applications or services are designed to suit students' preferences and tastes (Lefever and Currant, 2010). The effects are extremely positive when such a strategy is used in the context of remedial programmes or with students at risk as it can contribute to the struggle against dropping-out or school disaffection (Holley and Oliver, 2010).

Convenience

In the domain of technology, convenience can be defined as a satisfactory user experience that results from the provision of solutions that are suitable to user's comfort, purposes or needs. Quite often the discussions about the provision of technological solutions for education tend to focus on their relevance to learning goals. Rarely is this related to the learner's inner experience with the technological solution (Dede, 2005).

In education it is important that technological solutions add value to educational and/or administrative processes by making them more convenient to the learner. This is, for instance, what happens when online applications help students to continue their work out of school, anytime, anywhere, but also, more controversially, when they are allowed, if not encouraged, to do it collaboratively (Chen, Lambert and Guidry, 2010). Not surprisingly, many technology-based educational innovations can be said to have failed because instead of providing a much more convenient educational environment, they have required a much higher effort from the learner.

The two landmarks of learner's convenience in education are flexibility and personalisation. All in all, technology is seen as being suited to encouraging engagement by providing richer learning environments and flexibility, which is a key aspect of convenience (Joint Information Systems Committee, 2009; Williams and Chinn, 2009). Technology has the potential to help people understand and respond to students' needs better and to offer a more personalised experience (Harvey, 2006; Heaton-Shrestha, 2009). Initiatives are seen as being user-centred or user-convenient and, therefore, able to enhance a sense of empowerment by being flexible, personalised and focused on students generating knowledge and valuing their voice. The flexibility technology offers is viewed as enabling students to study according to their preferred way, in their own time and, therefore, offers them a sense of ownership and control (Heaton-Shrestha, 2007).

This "anytime, anywhere" learning links to the concept of providing personalised learning and to studies that view technology as able to offer control and autonomy to students by putting tools into their hands (Rismark, 2007). Technology can allow access to personalised services provided by the institution, but it can also offer the opportunity to use personal tools to suit individual needs. For example, online resources are perceived as offering more choice, and students match these to their personal preferences and use them to support their individual learning goals (Shroff and Vogel, 2009).

Enhanced productivity

Although the concept is unusual in educational literature, it is a fact that both teachers and students can obtain gains in productivity thanks to improved methodologies, and technology can help in this endeavour. Productivity can be defined as the relative efficiency of a person when performing a particular task. In education, increased productivity would mean an improvement in performance given the same amount of time spent on the task.

Recent OECD work on the New Millennium Learners has contributed to elucidating the role that technology plays in the educational productivity paradox seen from the perspective of the efficiency of teaching methodologies (OECD, 2010). As in other domains of human activity, technology use translates into productivity gains and increased efficiency only if working methods are transformed. In other words, there are limits to the learning gains that traditional education methods can expect from technology solutions; rather, there is a rational expectation that a change in education methodologies that benefits from technological change could result in significantly increased educational performance. In addition, this methodological change may be better suited to the skills and needs of the knowledge economy.

Students see in technology an opportunity to increase their productivity – even if only a few of them would use this term to describe their rational expectations. They expect technology solutions and, in particular, connectedness to either increase their educational performance or at least reduce the amount of effort invested in effective learning. If neither of these two goals are achieved, then a feeling of failure emerges.

Students' rational expectations regarding learning efficiency through technology are not without problems (Pedró, 2010). Firstly, students often express reluctance towards educational innovations that they may see simply as experiments without any direct incentives or benefits for them. This might be related to the uncertainty, disruptiveness and discomfort that discrete technology-based innovations may cause them. Secondly, their concept of "being efficient" may raise critical questions in their teachers. In particular, teachers may feel disappointed when realising the "productive and efficient" ways in which their students deal with assignments. Clearly, students will naturally use the tools they have at hand. Teachers cannot keep the same assignments that they designed well before their students could get connected; students will find ways to use connectedness to at least reduce the time on task, thus becoming much more efficient but with dubious learning gains. Third, students' familiarity with technology does not easily translate into an efficient use of connectedness in learning. Without adequate training and support in skills for learning, students may lose important opportunities to become much more learning efficient. Fourth, there is a remarkable number of "connectedness gaps" among students, and the productive paradox widens the range between the most and least connected (distances often based on gender, age, economic conditions and cultural divides).

Notes

1. More information about this project can be found at *www.elene-tlc.net.*

2. For a more detailed account see The New Millennium Learners: Learning From Learners' Voices [EDU/CERI/CD(2009)23].

3. For a comparative analysis of some of these surveys see Higher Education Academy (2007).

References

BellSouth Foundation (2003), *The Growing Technology Gap Between Schools and Students. Findings from the BellSouth Foundation Power to Teach Program*, BellSouth Foundation, Atlanta.

Bennett, S. and K. Maton (2010), "Beyond the 'digital natives' debate: towards a more nuanced understanding of students' technology experience", *Journal of Computer Assisted Learning, 26*, pp. 321-331.

Carey *et al.* (2009), *Privacy and integrity in the virtual campus*, paper presented at the Proceedings of network ethics: The new challenge in business, ICT and education conference, track 3: Higher education and virtual learning: Ethical issues and perspectives.

Caruso, J.B. and R.B. Kvavik (2005), *ECAR Study of Students and Information Technology, 2005: Convenience, Connection, Control, and Learning*, Educause Center for Applied Research, Washington, DC.

Caruso, J.B. and R.B. Kvavik (2006), *Preliminary Results of the 2006 ECAR Study of Students and Information Technology*, Educause Center for Applied Research, Washington, DC.

Crook, C. (2012), "The 'digital native' in context: tensions associated with importing Web 2.0 practices into the school setting", *Oxford Review of Education*, Vol. 38(1), 63-80.

Chen, P.-S.D., A.D. Lambert and K.R. Guidry (2010), "Engaging online learners: The impact of Web-based learning technology on college student engagement", *Computers and Education,* Vol. 54(4), 1222-1232.

Chung, C.J. and W. Storm (2010), *Using Interactive Technologies to Promote Student Engagement and Learning in Mathematics,* paper presented at the Proceedings of Society for Information Technology and Teacher Education International Conference 2010, Chesapeake, VA.

Dede, C. (2005), "Planning for neomillennial learning styles", *Educause Quarterly, 1*, 7-12.

Flather, P. and R. Huggins (2004), *Europaeum survey*, Future of European Universities Project, DaimlerChrysler Services AG, Oxford.

Hargittai, E. (2010), "Digital Na(t)ives? Variation in internet skills and uses among members of the 'Net Generation'", *Sociological Inquiry,* Vol. 80(1), pp. 92-113.

Harris *et al.*, (2010), "Small steps across the chasm: ideas for embedding a culture of open education in the university sector", *Education,* Vol. 16(1), *www.ineducation.ca/article/small-steps-across-chasm-ideas-embedding-culture-open-education-university-sector.*

Harvey, L. (2006), *The First Year Experience: a review of literature for the HEA academy*, Higher Education Academy, London.

Heaton-Shrestha, C. (2007), "Learning and E-Learning in HE: The Relationship between Student Learning Style and VLE Use", *Research Papers in Education,* Vol. 22(4), pp. 443-464.

Heaton-Shrestha, C. (2009), "Student Retention in Higher Education: What Role for Virtual Learning Environments?", *Journal of Further and Higher Education,* Vol. 33(1), pp. 83-92.

Higher Education Academy (2007), *Comparative review of British, American and Australian national surveys of undergraduate students*, *National Survey Comparative Review*, pp. 1-23.

Holley, D. and M. Oliver (2010), "Student engagement and blended learning: Portraits of risk", *Computers and Education,* Vol. 54(3), pp. 693-700.

Ipsos Mori. (2007), *Student Expectations Study. Key findings from online research and discussion evenings held in June 2007 for the Joint Information Systems Committee*, Joint Information Systems Committee (JISC), London.

Ipsos Mori. (2008), *Great Expectations of ICT. How Higher Education Institutions are measuring up*, Joint Information Systems Committee (JISC), London.

Joint Information Systems Committee (2009), *Higher Education in a web 2.0 world: Report of an independent committee of Inquiry into the impact on higher education of students' widespread use of web 2.0 technologies*, JISC, Bristol.

Jones, C. *et al.* (2010), "Net generation or Digital Natives: Is there a distinct new generation entering university?", *Computers and Education*, Vol. 54(6), pp. 722-732.

Kearsley, G. and B. Shneiderman (1998), "Engagement Theory: A Framework for Technology-Based Teaching and Learning", *Educational Technology,* Vol. 38(5), pp. 20-23.

Kennedy, G. *et al.* (2008), "First year students' experiences with technology: Are they really digital natives?" *Australasian Journal of Educational Technology,* Vol. 24(1), pp. 108-122.

Lam, I. and M. Ritzen (2008), *The ne(x)t generation students: needs and expectations*, Utrecht University, IVLOS – Institute of Education, Centre for ICT in Education, Bremen.

Lefever, R. and B. Currant (2010), *How can technology be used to improve the learner experience at points of transition?*, University of Bradford, Bradford.

Lenhart, A. *et al.* (2007), *Teens and Social Media. The use of social media gains a greater foothold in teen life as they embrace the conversational nature of interactive online media*, Pew Internet and American Life Project, Washington, DC.

Littlejohn, A., A. Margaryan and G. Vojt (2010), "Exploring students' use of ICT and expectations of learning methods", *Electronic Journal of E-Learning (IJEL),* Vol. 8(1), *www.ejel.org/Volume-8/v8-i1/v8-i1-art-2.htm.*

Margaryan, A., A. Littlejohn and G. Vojt (2010), "Are digital natives a myth or reality? University students' use of digital Technologies", *Computers and Education*, Vol. 56(2), pp. 429-440.

Nagler, W. and M. Ebner (2009), *Is your university ready for the Ne(x)t-Generation?*, paper presented at the Proceedings of 21st world conference on educational multimedia, hypermedia and telecommunications (EDMEDIA), Honolulu.

Newmann, F. (1986), "Priorities for the future: Toward a common agenda", *Social Education,* Vol. 50(4), pp. 240-250.

Oblinger, D.G. and B.L. Hawkins (2005), "The myth about E-learning", *Educause review,* Vol. 40(4), pp. 14-15.

OECD (2005), *Are students ready for a technology-rich world? What PISA studies tell us*, OECD, Paris.

OECD (2010), *Are the New Millennium Learners Making the Grade? Technology Use and Educational Performance in PISA*, OECD Publishing, Paris.

Pedró, F. (2010), "ICT and Postgraduate Education", in T. Kerry (ed.), *Meeting the Challenges of Change in Postgraduate Education* (pp. 105-121), Symposium Books, London/New York.

Rismark, M. (2007), "Using Mobile Phones to Prepare for University Lectures: Student's Experiences", *Turkish Online Journal of Distance Education*, Vol. 6(4), article 9.

Shroff, R.H. and D.R. Vogel (2009), "Assessing the Factors Deemed to Support Individual Student Intrinsic Motivation in Technology Supported Online and Face-to-Face Discussions", *Journal of Information Technology Education* (8), pp. 11-22.

Spot+ (2002), *Survey report: Students' perceptions of the use of ICT in university learning and teaching*, The SOCRATES Programme – MINERVA Action, Brussels.

Trigwell, K., M. Prosser and F. Waterhouse (1999), "Relations between teachers' approaches to teaching and students' approaches to learning", *Higher Education,* Vol. 37(1), pp. 57-70.

Virtanen, V. and Lindblom-Ylänne, S. (2009), "University students' and teachers' conceptions of teaching and learning in the biosciences", *Instructional Science,* Vol. 38(3), pp. 355-370.

Williams, J. and S.J. Chinn (2009), "Using Web 2.0 to Support the Active Learning Experience", *Journal of Information Systems Education,* Vol. 20(2), pp. 165-174.

Zemsky, R. and W.F. Massy (2004), *Thwarted Innovation, What happened to e-learning and why,* The Learning Alliance at the University of Pennsylvania, Philadelphia.

Chapter 6

Emerging issues for education

The empowerment of children and young people through digital media and connectivity has resulted in a number of new challenges to education. Very little is known about the effects of becoming a content producer with a potentially unlimited audience at a very early age. Even less about the impact of creating and nurturing virtual social networks which are, in many ways, free from any adult supervision. These emerging issues in relation to how young people use digital media and connectivity can be mapped out in five different domains: entertainment, information, knowledge and learning, social and psychological. Although the resulting issues can be seen as sources of concern, they can also be regarded in most cases as opportunities for engaging in an educational dialogue with learners.

Throughout the previous two chapters the issues of the effects of connectivity and media attachment on learners and on the eventual changes in their learning expectations have been addressed from an evidence-based perspective. Yet, there are a number of other concerns that should matter both to policy makers and educators. All of them stem from the fact that education has the responsibility to equip young people with the necessary skills and values that will allow them to cope with the challenges that connectedness is currently posing to them. All of them can be considered as emerging issues for education in the particular domain of technology and connectedness.

In this context, this section addresses the most salient issues until now and provides some analytical insights to contribute to a better understanding of the vast range of challenges now being faced by education. As has been often said (Bryant and Oliver, 2009), reflection upon these issues has only just started and it is far from having reached a consensus among the educational community. Therefore, what follows has to be taken more as an agenda-setting exercise rather than as a comprehensive set of validated responses.

Mapping out the issues

The current understanding of how technology use and connectedness affect learners and the research base that supports it are mostly based in a concept of learners as users or consumers of applications and services. However, in the last five years or so a new generation of applications and web-based services has emerged with the common denominator of allowing users to become producers and to create virtual identities which allow them to engage in a number of social spaces and activities.

Although, for instance, there is much hype about the Web 2.0[1] and its educational potential, very little is known about the effects of becoming a content producer with a potentially unlimited audience at a very early age and even less about the impact of creating and nurturing virtual social networks, be that through the or by way of using mobile phones and a particular idiom to set up communities which, in many ways, operate free from any adult supervision.

Although the range of issues that educators have now to address because of connectedness would be endless, Table 6.1 presents at least the main areas where concerns have been identified from an international perspective. For easy reference, they have been grouped into five different categories: entertainment, information, knowledge and learning, social, and psychological. Although all of them can be seen as sources of concern, they can also be regarded in most cases as sources of educational opportunity and in particular for engaging in an educational dialogue with learners.

Table 6.1. **Categories of issues facing educators due to increased connectedness**

Entertainment	Information	Knowledge and learning	Social	Psychological
• Videogames risks • Positive effects	• Information freedom and overload	• Digital literacy • Plagiarism • Clash of culture • Poorness of sms/chat language	• Safety • Different experiences of socialisation • Cyber-bullism • New ways to live the time	• Youth emancipation • Self-building

Entertainment

When it comes to digital media, entertainment is closely linked to video games. The challenges that video games pose to education can be seen both from a positive and negative perspective. Under the latter, video games are an open window to all kinds of risks, ranging from addiction to the development of aggressive if not violent behaviours. Yet, under a more positive approach video games are often said to be by their very nature a source of inspiration for learning with a potential that has not been fully explored yet in formal education settings.

The risks of video games

Video games are a fast changing source of entertainment that has been recently enhanced through online video games (including within social networking platforms) and through mobile devices. It is quite often said that the main risks associated with video games are the propensity to develop violent behaviours and addiction.

Violence in and addiction to video games are well known and much debated issues. Yet, there is no evidence as of yet of any relationship between playing very immersive video games, such as MMORPG (massively multiplayer online role-playing games) and violence (Min Lee, Peng and Park, 2009). The problem of addiction is more controversial and the discourse is made more complex by the influence of contextual and psychological factors and variables. Theoretical models to interpret these phenomena have been developed and, according to most of them, violent or addictive behaviours do not develop with video gaming as the only cause; rather video game exacerbates behaviours whose origins lie in personal situations often marked by a pre-existing psychological problem (Buckley and Anderson, 2006; Min Lee and Ping, 2006). A similar conclusion can be adopted in relation to addiction. Just to give an indication of how important addiction to video games, is it is worth mentioning research by Harris Interactive (2007), according to which 8.5 % of game users in the United States can be classified as addicts. These results are likely to be explained by psycho-sociological aspects, mainly: "social marginalization, emotional loneliness, and maladaptation to real life". Furthermore, it has been observed, "people of high physical aggression personality tend to play the game in a more violent way" (Min Lee *et al.*, 2009).

Positive effects on cognitive skills

Video games have been gaining increasing attention by educational experts in recent years. On the one hand, they are said to be models of the kind of engaging, immersive and well-planned learning journeys that should be mainstream in the 21st century (Miller, 2010). This was already anticipated years ago in a seminal work by James Paul Gee (2004) under the title of *What Video Games Have to Teach Us About Learning and Literacy.* On the other hand, the complex scenarios and situations, as well as simulation contexts, that characterise most video games, including the best-selling titles, may be seen as challenging for the player and potentially leading him to learn, even unconsciously, second order skills such as problem solving or evidence gathering (Shaffer, 2008). This broad assumption has led to the emergence of a dedicated category of video games for education and training purposes, often known as *serious* video games (Kankaanranta and Neittaanmäki, 2010). Yet, the investments required to produce high quality video games have limited their commercial development mostly to business or technical education.

Video gaming, no matter the type or theme of the game, has been found to have some positive side-effects. For instance, a number of researchers (Durkin, 2006; Min Lee *et al.*, 2009; Prensky, 2006; 2010) have found empirical evidence of a relationship between video gaming and the development or improvement of the following skills: strategising, problem-solving, recursive thinking, organisation of information in a given context, interpretation of visual information, heuristic ability, visualisation-spatial skills, mental rotation skills, and hand-eye coordination.

Information overload?

Another open-ended question raised by connectedness is related to information management. In particular it is worth considering the problem of the almost infinite amount of data available in the web that may easily lead to cognitive overload (Klingberg, 2008) or, alternatively, to oversimplification. In addition, the user has to be able to adopt a critical perspective on what they find in the Internet, which is a problem in turn linked to the credibility of sources.

By its very nature digital information is flexible, mutable, and in constant change – at least, this is the impression of any Internet surfer. The problem is that it is difficult to discern which pieces of information are valuable and trustworthy without some guidance. This creates a problem in terms of accountability: both governments and newspapers have to deal with it (Lin, 2009) as do, increasingly, individuals. This is likely to be the case for any child or adolescent who starts an Internet-based search without any previous guidance or training: what comes first is most convenient and thus the most likely to be used. An adolescent, on the Internet must be guided in their navigation, offered conceptual tools and equipped with critical information management skills. The main risk of information overloading is the feeling of saturation, which must be strongly avoided in order to help tomorrow's citizens to grow (Schwartz, 2005).

Knowledge and learning

Educators may find some of the issues that current behaviours inspired by technology tend to raise particularly challenging. Yet, they have to develop professional responses to each of them, namely what kind of digital literacy is needed to actively engage in the knowledge economy and society; how to address the inevitable problems posed by the fast copy and paste culture, and the implicit plagiarism; whether school and social practices by young people are always bound to produce a culture clash; and, finally, whether the new idioms developed by young users have to be banned or re-conducted.

Digital literacy

Conceptually, digital skills may be classified as instrumental (or basic or functional), informational (understanding, navigation, evaluation) and social (communication, self-disclosure, privacy). According to recent research (Sonck *et al.*, 2011), all these skills are all correlated with each other, forming a single scale. Thus, it seems that those who are able to judge the veracity of websites are also those who can find safety information, those who can bookmark a site can also block unwanted messages. On the other hand, those who struggle with one skill are likely to struggle with another. Thus, the more experienced the user becomes, the wider the range of skills she develops.

Yet, younger children, girls and those from lower SES homes are gaining fewer skills (because they do less online, for various possible reasons). In particular, low skills among 11-13 year olds poses a challenge for teachers, parents and others. Fewer than half can block unwelcome messages or find safety information or bookmark a site, and only a third can compare websites to decide if information is true or block unwanted junk mail. By trial and error or peer-to-peer learning, older children do gain digital skills, including safety skills. But as ever younger children go online it is increasingly important to raise awareness among educators, including parents, about the importance of starting early with digital literacy.

Digital literacy is not meant to replace literacy as it has traditionally been understood, but has to be integrated with concurrent forms of literacy and particularly with traditional paper and pencil literacy. 21st century skills have not evolved from previous skills, but they are a set of new competencies. In fact, there are multiple layers of skills and competencies populating the relationship linking ICT innovation, actions using digital technologies, and thinking and learning development. All this suggests that living in the digital era implies that one has to be digitally educated, building a constantly growing set of skills that has to be closely related to everlasting educational goals (McCannon, 2009; Media Awareness Network, 2010).

Yet, the problem is that digital literacy tends to be marginalised from schools (Thomas, 2011). Despite the many studies suggesting that without a direct instructive intervention digital literacy hardly develops (Carrington and Robinson, 2009; Selwyn, 2010, 2011), few curriculum reforms seem to have taken this into account seriously, that is, translating statements into requirements for national student assessments.

Plagiarism and the "copy-paste culture"

The many and easy ways in which information can be found on the Internet and copied and pasted has also raised concerns about plagiarism, particularly in university settings but also, increasingly, in schools. This is, in fact, an indication of the shortcomings of the lack of appropriate media education in schools, and the need to incorporate in curricula not only the technological skills required to manage information from a technical point of view, but also the values that inspire concepts which are difficult to grasp at early ages but may have a long lasting impact, such as intellectual property, academic authority, or even the difference between finding and downloading information and constructing knowledge, personally or collectively.

As Nilsson (2008) found in a pioneering research about students' attitudes towards plagiarism, it is important to distinguish between actions and the meaning ascribed to them as acts by students – that is, whether they are really aware of the moral values involved in their current practices with knowledge production as learners. Buckingham (2006), for instance, has suggested that the analytical framework of media literacy might be successfully applied in schools as a tool for developing consciousness and reflective practices when using the Internet. Moreover, Jenkins *et al.* (2007) claim that an important goal of media education should be to encourage young people to become more reflective about the ethical choices they make as participants and communicators and the impact they might have upon others.

Clash of cultures

Lastly, there is the issue of whether connectedness and the tools to deal with digital media which empower children and young people as content creators may in the long run affect their expectations as learners and their behaviours in classrooms. Some would claim that because of this, pupils are increasingly challenging teachers' and parents' authority, and that the virtual communities in which they participate make them conform to values, rules and norms that defy those traditionally heralded in schools. From this perspective, Jenkins *et al.* (2007) highlight the need to ensure that every child has been socialised into the emerging ethical standards which should shape their practices as media makers and as participants within online communities. New participatory culture places new emphasis upon familiar skills as well as requires teachers to pay greater attention to the social skills and cultural competencies that are emerging in response to changes in the media landscape.

As Kolikant has put it (Kolikant, 2010), it may be that today's students live within two value systems regarding the Internet, computers, ICT, etc., and their usefulness and appropriateness for learning purposes. Outside school, students are used to interacting with information through collaboration and bricolage:[2] "Today's kids get on the web and link, lurk, and watch how other people are doing things, then try it themselves" (Brown, 2000, p. 14). "Produsaging"[3] (Bonk, 2009) and sharing are typical too and considered acceptable by ICT users. They can legitimately "find something – an object, tool, document, a piece of code – and [use] it to build something [they] deem important." (Brown, 2000, p. 14). For example, they can prepare (and upload) a summary about a certain topic taken from other summaries downloaded from the Internet. Thus, the Internet encourages and facilitates a culture of sharing. Simply attempting to hold information in your head is not valued but being able to navigate and find what is needed and meliorate it, *e.g.* use it in innovative ways is (Passig, 2003, 2007). In contrast, in-school students encounter "person-solo" learning (Perkins, 1992), operating under the assumption that knowing means having knowledge in one's head (Shaffer and Clinton, 2006). Therefore, a partnership with others and/or information technology is legitimate, but only for the sake of the learning process, *i.e.* the process of knowledge acquisition (or building). Exams, important events in students' lives, demonstrate these assumptions. In the classes studied, as well as in many other schools, students are most commonly assessed individually with closed books, and certainly not with the Internet. Another example of this value system is the long debate about whether, when and how to allow the use of calculators in math classes (Dick, 1988; Ellington, 2003; Reynolds, 1993).

Yet, it is important to examine this clash of cultures in terms of their impact on learning. Existing empirical research about the relationships between the use of social networking sites (such as Facebook) and educational performance is still in a very early phase. There is still some ground to cover before the use of Web 2.0 in the teaching and learning processes is close to reaching critical mass, which would enable substantial effects on teaching and learning (Schmidt and Vandewater, 2008). Today, the results seem to point to a clear relationship between this use and lower academic performance, although no causal relationships have been established (Eberhardt, 2007; Kirschner and Karpinski, 2010; Madge, 2009; Selwyn, 2009).

Clearly, technology and connectedness also offer a unique opportunity to bridge the gap between these two cultures. Since both technology and connectedness are already part of the daily experience of most students outside schools, both of them represent a window of opportunity for enhanced anytime, anywhere learning. Schools may gain a lot by using this opportunity to bridge the gap.

The new idioms

Connectedness also challenges the traditional patterns of language use not only in informal communication but beyond. As a matter of fact, some particular characteristics of the medium used, be that sms, chat or a social networking site, have contributed to the emergence of new idioms. All these idioms tend to generate quick forms of expression that are only marginally related to the canon. They all intend to replicate oral communication. Moving from that need, two effects emerge: the diffused practice to colour digital messages (sms, chat, microblogging, etc) with emoticons; the growing of a new sort of language (half-oral half-written) meant to shorten as much as possible the time spent in writing.

Since this evolution of the language is happening all around the world, and teachers report that they have started to notice it in school texts, it is important to find a criterion to discriminate within such a linguistic challenge. Main concerns of linguists come from the perpetual issue: given that a language is a free and flexible entity, how to manage the contrast between new kinds of communication and loss of grammatical skills (Campbell and Ling, 2009).

Socialisation using digital tools

The fact that connectedness puts young people in connection with others (of varying ages) raises a number of concerns and fears. The first is about safety. The second one relates to bullying. A third concern is related to the different experiences of socialisation that connectedness can provide. A final concern lies precisely in the implications that such an intensive social experience may have over the management of time.

Safety

Children have the right to protection and safety online but they must also take responsibility for keeping safe and respecting the rights of others on the Internet. Young people's access to the Internet is increasingly less open to adult supervision. Nearly half of all children in Europe go online in their own bedroom (O'Neill, Livingstone and McLaughlin, 2011) where it is unrealistic to expect parents to monitor their safety. As a recent OECD (2011) report suggests, three broad categories of online risks for children can be considered: *i)* content and contact risks, including exposure to pornography, cybergrooming and cyberbullying; *ii)* consumer risks related, for example, to online marketing and fraudulent transactions; and *iii)* privacy and security risks, including the use of social networks without sufficient understanding of potential long-term consequences. While the same spectrum of risks is present in all countries, the available data suggest that prevalence rates vary. Moreover, because children's activities, skills and resilience differ, their interactions with the online environment and the consequences differ as well. While children's capabilities are likely to increase with age, so can their own risky behaviour.

Safety in virtual spaces has been receiving increasing attention by the media. Even technology providers have developed a number of solutions intended to prevent access to inappropriate content, some of them to be used at home while others are for schools and universities. All these solutions offer tools that are intended to enforce adults' supervision and to let them decide what is appropriate for a child or a student to look at on the Internet. But neither of them can be seen as substitutes for parental and educational responsibilities over children's upbringing. Both the generalisation of the use of cellular phones, with devices intended for children as young as 8 years old, and the widespread and viral penetration of social applications on the Internet allow them to generate spaces whose rules, contents, inner life and, most importantly, actual members, some of them with faked identities, completely escape adult detection and responsible supervision.

Online risks faced by children are many and evolving. Addressing them requires a blend of approaches that include legislative, self- and co-regulatory, technical, awareness and educational measures, as well as positive content provision and child safety zones. In practice, each country operates its own policy mix of characteristics and priorities, which reflects its perception of priorities as well as its culture and style of government. Moreover, policy measures that address different risks and initiatives from various stakeholders at different levels co-exist. This creates policy complexity at national level and policy heterogeneity across countries.

Yet, there is growing evidence showing that young people are developing some self-defence tools and, on average, they are aware of the many risks of connectedness. As stated in a recent report: "Research suggests that popular narratives around sexual predators on SNSs are misleading – cases of unsuspecting teens being lured by sexual predators are rare … Furthermore, only .08% of students surveyed by the National School Boards Association (2007) in the United States met someone in person from an online encounter without permission from a parent" (Rideout *et al.*, 2010). Yet, with the average age of first Internet use at 7 in countries such as Denmark and Sweden, and 9 overall throughout Europe (O'Neill *et al.*, 2011), there needs to be a new policy focus on much younger children for whom the Internet is an everyday experience and a greater concentration of effort at primary school and even pre-school level.

Across the 25 countries surveyed by EU Kids Online, less than one third (28%) of parents were found to filter the websites visited by their child. But trends are not uniform across countries. For example, according to Marwick *et al.* (Marwick, Murgia-Diaz and Palfrey, 2010) who compared studies carried out in 2005, 2007 and 2009, the use of filtering software in the United States increased from 44% to 56%. However, in the United Kingdom, Ofcom found a decrease in the use of control or filtering software by parents, from 49% in 2008 to 43% in 2009 (Ofcom, 2010). It is clear that many parents find such software either too complicated or ill-suited to their needs. To be effective, parental controls need to incorporate all of the issues that concern parents about their children's Internet use. Thus, in addition to filtering out adult or unsuitable online content for children, controls may also need to include features such as the amount of time spent online, filtering of user-generated content and blocking of commercial content.

Awareness-raising should emphasise empowerment rather than restriction, and appropriate, responsible behaviour with regard to technology use. On top of technical tools, what matters is the ability of parents to become a valuable source of advice and education in this domain. Digital parenting involves encouraging children and young people to develop self-governing behaviour in which they take greater responsibility for their own safety in the use of the Internet. As 9% of 9-10 year olds have been bothered or upset by something on the Internet in the past year, it is important to promote awareness-raising and other safety practices for ever younger children.

Cyber-bullying

Among the new experiences of online socialisation, cyber-bullying has become an educative concern. Researchers and observers do not agree on the nature of the problem and it is possible to find definitions built around the idea of a new type of bullying and other views, suggesting that flaming online and blaming face-to-face are not so different. The debate in educational terms surrounds which kinds of counter-moves are likely to be effective; an interesting approach proposed by Lange turns around the education to responsibility (Lange, 2007).

Different experiences of socialisation

If the safety issue can be considered an overstated concern, there are fewer outstanding problems that can have a big effect on socialisation dynamics. The most reported is the gap between the – very often – uncontrolled and overdrawn self-disclosure on the web and the control over privacy of information; adolescents express a sort of split between declared awareness and acted behaviours (Christofides, Muise and Desmarais, 2009). Shyness is often reported as a sensible variable in a range of factors: the willingness to self-project and the fear of others' judgments merge in an online hyper-active, social arena in which the rules of socialisation are less clear, where it is possible for young students to control technical protective barriers themselves (Huang and Leung, 2009). The job market is taking advantage of all the information spread across the Internet by young people (Bohnert and Ross, 2010), and it is common for interviewers to study the online profiles of candidates prior to meeting them. The contrast between the carelessness applied in updating a web-profile, and the seriousness of its effects must be considered in media literacy programmes.

Both the Internet and new mobile devices offer new kinds of interaction, but above all, Social Networking Sites (SNS) multiply online socialisation, and create new issues to be addressed. Loneliness and hyper-interaction could be the two sides of the same coin (Lin, 2009), and the absence of adult supervision increases the risk posed to students. It has been observed that SNS develop around niches of interest and grow thanks to the involvement of "friends of friends" and the increasing overlap of networks within the Internet (Rideout *et al.*, 2010). Given this new way to socialise, the concept of a cohort changes dramatically, and the peer group can grow without the traditional limits of age and space.

A new way to spend time

Merging the potential of smartphones with the amount of information related to emailing and social networking, the distinction between public and private moments has been reduced, as well as the experience of (potentially) never-ending connectedness for people in OECD countries (Campbell and Ling, 2009; Lin, 2009). Recent studies are observing how the contemporary way to live in and perceive time does not affect only the time spent connected, but also face-to-face experiences. In this regard the "loneliness of mobile life" has been explored (Bardi and Brady, 2010). This kind of social behaviour can be recognised, for instance, when people do not feel at ease in face-to-face interactions and take refuge in typing a message, checking emails, or (even more contradictory) updating their web-profile to express personal discontent globally.

Clearly, here, the challenge is to imagine ways to (self-)regulate connectedness in order to get the best out of digital socialisation, learning how to set the media diet with the goal of regulating how one spends their time, and avoiding the use of mobile technology to avoid face-to-face socialisation. NML need to be helped in this process and shown both the pros and the cons of being alone, so that they can choose consciously when it is worthwhile to be connected and when it is not.

Another relevant aspect of being connected is the high time-consumption it requires, very often without defined perception. Young people are increasingly going online just for fun and to pass the time. On any given day, 53% of all the young adults aged 18-29 in the United States go online for no particular reason except to have fun or to pass the time (Pew Research Center's Internet and American Life Project, 2011). Young adults' use of the Internet can at times be simply for the diversion it presents. Indeed, 81% of all young adults in this age cohort report they have used the Internet for this reason at least occasionally. Studies on media diet report that, versus many hours spent online, respondents estimated that they were connected for much less time; this is mainly due to the coexistence of multiple tasks and to the dramatic underestimation of the time devoted to communicating through digital devices and online facilities (Rideout *et al.*, 2010). In terms of learning, this issue has been related to ADD (Attention Deficit Disorder) (Crenshaw, 2008). Even if there is no indubitable evidence about the negative effects of connectedness, it is worth taking into serious consideration the risks of underestimating Internet addiction (Byun *et al.*, 2009; Yen *et al.*, 2009), understandable as "it gives you the feelings and gratifying sensations that you are not able to get in other ways. It may block out sensations of pain, uncertainty or discomfort. It may create powerful distracting sensations that focus and absorb attention. It may enable a person to forget or feel 'okay' about some insurmountable problems. It may provide an artificial, temporary feeling of security or calm, of self-worth or accomplishment, of power and control, or intimacy or belonging" (Young, 1999).

This definition must not be considered literally, but it is true that some of these feelings are typical of hyper-connected people. It is remarkable that, even when the goal is marketing, social networking seems to be inappropriate in terms of return on investment, because of the large amount of time devoted to it (Stelzner, 2010).

Psychological issues: new problems or new opportunities?

Irrespective of which technologies young people may be using, the fact is that young people are, by definition, in a process of building their own identity, which quite often means developing some form of transitional emancipation from the adults surrounding them. Both in relation to emancipation and to identity building, connectedness is bound to raise a number of concerns for educators.

Youth emancipation through connectedness

Thanks to the Internet, young people's curiosity can be fed – potentially – whenever, on whatever. Opportunities offered by the digital context are both a stimulus and a highway to knowledge; if compared with previous generations, NML are likely to achieve intellectual independence from their educators at earlier stages of life, due to the access to information and to the enhanced style of retrieving it, by surfing. It is arguable that these ways of learning are more exciting than parental or school transmission of knowledge; therefore it must be welcomed that young people can navigate their paths of interests with freedom (Campbell and Ling, 2009). The issue that remains unresolved is the risk of improving digital orphanage. Processes of emancipation need to be in the context of boundaries, education and feedback regarding limiting connectedness and personal psychological development. While the effect of accessing information and socialisation helps emancipation by offering many and different stimuli, it must be paired with the reassuring presence of responsible adults, able to be a mirror for the projection of "self".

Identity building

The development of self is a very complex and delicate part of life, especially during adolescence (when NML are more exposed to connectedness) and it is crucial to build an identity that is aware, independent, and solid. What has been described in this section represents a great challenge. If one only considers the impact of mobile and social networking on the everyday lives of NML it is evident that every opportunity can also pose a risk. Younger people are asked to deal with a variety of experiences, which can lead them to digital schizophrenia and to stressful self-representation (Buckingham and Willett, 2006). The possibility of reshaping intimacy, privacy and self-expression can be a great influence on self-building, or can lead to uncertainty of identity (building, deconstructing and reconstructing online profiles, playing different roles face-to-face and online) (Livingstone, 2008). The complexity of this subject is clear when one considers how important it is to a student that the image they project, by way of personal pictures online for instance, is accepted by their peer group and social network (Boyd, 2008).

Notes

1. As opposed to the original web (version 1.0) where users mostly searched and downloaded information, Web 2.0 emphasises the role of the user as an active content-producer: people are expected to upload information, sometimes personal, share it with others, or engage in multiple conversations in a variety of ways ranging from blogs and wikis (such as the Wikipedia) to social applications intended to create and maintain communities with shared interests, values or links. The term was originally coined by Tim O'Reilly (2005).

2. Bricolage, or tinkering, "can be taken to mean 'trial-and-error,' learning by poking around, trying this or that until you eventually figure it out" (Papert, 1996). Bricolage is also about "the abilities to find something – an object, tool, document, a piece of code – and to use it to build something you deem important" (Brown, 2000, p. 14).

3. "Produsaging" is a new hybrid form of simultaneous production and usage amidst connected people (Bruns, 2007). Produsagers are users engaged in collaborative and continuous building and extending of existing content in pursuit of further improvement (*e.g.* Wikipedia).

References

Bardi, C. A. and Brady, M. F. (2010), "Why shy people use instant messaging: Loneliness and other motives", *Computers in Human Behavior,* Vol. 26, pp. 1722-1726.

Bohnert, D. and Ross W.H. (2010), "The Influence of Social Networking Web Sites on the Evaluation of Job Candidates", *Cyberpsychology, Behavior, and Social Networking,* Vol. 13, pp. 341-347.

Bonk, C. (2009), *The world is open: how Web technology is revolutionizing education*, Jossey-Bass, San Francisco.

Boyd, D. (2008), "Why Youth (Heart) Social Network Sites: The Role of Networked Publics in Teenage Scoial Life", *Youth, Identify, and Digital Media*, the John D. and Catherine T. MacArthur Foundation Series on Digital Media and Learning, David Buckingham (ed.), MIT Press, Cambridge, MA.

Brown, J.S. (2000), "Growing up digital: How the Web changes work, education, and the ways people learn", *Change , Vol.* March/April, pp. 10-20.

Bruns, A. (2007a), "Produsage, Generation C, and Their Effects on the Democratic Process", MiT *5 (Media in Transition) conference*, MIT, Boston, USA.

Bruns, A. (2007b), *Produsage : Towards a Broader Framework for User-Led Content Creation, in Creativity and Cognition : Proceedings of the 6th ACM SIGCHI conference on Creativity & Cognition*, ACM, Washington, DC

Bryant, J. and M. B. Olivier (2009), *Media effects: advances in theory and research*, Routledge, New York.

Buckingham, D. (2006), "Defining digital literacy. What do young people need to know about digital media?" *Digital kompetanse, Nordic Journal of Digital Literacy,* Vol. 1(4), pp.263-276.

Buckingham, D. and R. Willett (2006), *Digital generations: children, young people, and new media*, Lawrence Erlbaum Associates Publishers.

Buckley, K.E. and C.A. Anderson (2009), "A theoretical model of the effects and consequences of playing video games" in P. Vorderer and J. Bryant (eds.), *Playing Video Games- Motives, Responses, and Consequences,* Routledge, New York.

Byun, S., *et al.* (2009), *Internet addiction: Metasynthesis of 1996-2006 quantitative research*, CyberPsychology and Behavior.

Christofides, E., Muise, A., and Desmarais, S. (2009), "Information disclosure and control on Facebook: Are they two sides of the same coin or two different processes?", CyberPsychology *and Behavior.*

Crenshaw, D. (2008), *"The Myth of Multitasking: How 'Doing It All' Gets Nothing Done",* Jossey-Bass, San Francisco.

Dick, T. (1988), "The continuing calculator controversy", *Arithmetic Teacher*, Vol. 35(8), pp. 37-41.

Durkin, K. (2009), *Game playing and adolescents' development,* in P. Vorderer & J. Bryant (eds.), *Playing Video Games: Motives, Responses, and Consequences,* pp. 491-507.

Eberhardt, D.M. (2007), "Facing Up to Facebook", *About Campus*, Vol. 12(4), pp. 18-26.

Ellington, A.E. (2003), "A meta-analysis of the effects of calculators on students' achievement and attitude levels in precollege mathematics classes", *Journal for Research in Mathematics Education,* Vol. 34(5), pp. 433-463.

Gee, J.P. (2004), *What Video Games Have to Teach Us About Learning and Literacy*, Palgrave Macmillan, New York.

Harris Interactive (2007), "Video Games Addiction: Is it Real?", Rochester N. Y, *www.harrisinteractive.com/news/newsletters_k12.asp.*

Huang H. and L. Leung (2009), "Instant messaging addiction among teenagers in China: Shyness, alienation, and academic performance decrement", *Cyberpsychology and Behavior*, Vol. 12, no. 6.

Jenkins, H. *et al.* (2007), "Confronting the Challenges of Participatory Culture: Media Education for the 21st Century", *Digital Kompetanse. Nordic Journal of Digital Literacy*, Vol. 2(1).

Kankaanranta, M.H. and P. Neittaanmäki (eds.) (2010), *Design and Use of Serious Games*, Springer, New York.

Kirschner, P.A. and A.C. Karpinski (2010), "Facebook and academic performance", *Computers in Human Behavior, 26*, pp. 1237-1245.

Kolikant, Y.B.-D. (2010), "Digital natives, better learners? Students' beliefs about how the Internet influenced their ability to learn", *Computers in Human Behavior, 26*, pp. 1384-1391.

Lange, P. (2007), "Publicly private and privately public: Social networking on YouTube", *Journal of Computer-Mediated Communication*, Vol. 13(1).

Lee, K., Peng, W. and Park, N. (2009), "Effects of computer/video games, and beyond", in J. Bryant, and M. B. Oliver (eds.), *Media Effects: Advances in Theory and Research,* Routledge, New York.

Lin, Y.R., *et al.,* (2009), "Analyzing Communities and Their Evolutions in Dynamic Social Networks. ACM Transactions on Knowledge Discovery from Data (TKDD)", special issue on *Social Computing, Behavioral Modeling, and Prediction*, Vol.3, Issue 2, pp.1-3.

Ling, R.S. and S.W.Campbell (2009), *The Reconstruction of Space and Time: Mobile Communication Practice*, Transaction Publishers, New Brunswick, N.J.

Livingstone, S. (2008), "Taking risky opportunities in youthful content creation: teenagers' use of social networking sites for intimacy, privacy and self-expression", *New Media & Society*, NSBA, *Vol.10, pp.* 393-411

Madge, C. (2009), "Facebook, social integration and informal learning at university: 'It is more for socialising and talking to friends about work than for actually doing work'", *Learning, Media and Technology,* Vol. 34(2), pp. 141-155.

Mahwah, N. J. Ling, R. and S.W. Campbell (eds.) (2009), *The reconstruction of space and time: Mobile communication practice*s, Transaction Publishers, New Brunswick, NJ.

Marwick, A., D. Murgia-Diaz, and J. Palfrey (2010), *Youth, Privacy and Reputation (Literature Review)*, Vol. 2010-5, Berkman Center Research Publication, New York.

McCannon, D., 2009, "Visualising the Essay - Using information graphics to facilitate critical thinking within an art school", Third Global Conference in Visual Literacies.

Media Awareness Network (2010), "Digital Literacy in Canada: From Inclusion to Transformation", *www.media-awareness.ca/english/corporate/media_kit/digital_literacy_paper_pdf/digitalliteracypaper.pdf.*

Miller, C.T. (2010), *Games: Purpose and Potential in Education*, Springer, New York.

Min Lee, K., and W. Peng (2009), "What do we know about social and psychological effects of computer games? A comprehensive review of the current literature", in P. Vorderer, J. and Bryant (eds.), *Playing video games: Motives, responses, and consequences,* pp. 383-407), Routledge, New York.

Min Lee, K., Peng, W. and Park, N. (2009), Effects of computer/video games, and beyond, J. Bryant and M. B. Oliver (Eds.), *Media Effects: Advances in Theory and Research,* pp. 551-566, Routledge, New York.

National School Boards Association (NSBA) (2007), *Creating and Connecting*, Alexandria.

Nilsson, L.E. (2008), "But Can't You See They are Lying?" *Student moral Positions and Ethical Practices in the Wake of Technological Change*, University of Gothenburg, Department of Education, Sweden.

O'Reilly, T. (2005), "What Is Web 2.0?", retrieved from *www.oreillynet.com/pub/a/oreilly/tim/news/2005/09/30/what-is-web-20.html.*

Office of Communications (Ofcom) (2010), "UK children's media literacy", *http://stakeholders.ofcom.org.uk/binaries/research/media-literacy/ukchildrensml1.pdf*

OECD (Organisation for Economic Co-ordination and Development) (2011), "The Protection of Children Online: Risks Faced by Children Online and Policies to Protect Them", OECD Digital Economy Papers, No. 179, OECD Publishing, Paris.

Papert, S. (1996), *The connected family: Bridging the digital generation gap*, Longstreet Press, Atlanta.

Passig, D. (2003), "A taxonomy of future thinking skills", *Informatics in Education,* Vol. 2(1), pp. 79-92.

Passig, D. (2007), "Melioration as a higher thinking skill of future intelligence", *Teacher College Record,* Vol. 109(1), pp. 24-50.

Perkins, D.N. (1992), *Smart schools: From training memories to educating minds*, Free Press, New York.

Pew Research Center's Internet and American Life Project (2011), *The internet as a diversion and destination*, Pew Research Center, Washington, DC.

Prensky, M (2006), *Don't bother me Mom, I'm learning : How computer and video games are preparing your kids for twenty-first century success and how you can help!*, Paragon House, St. Paul, Minn., USA.

Reynolds, B.E. (1993), "The algorists vs. the abacists: an ancient controversy on the use of calculators", *The College Mathematics Journal,* Vol. 24(3), pp. 218-223.

Schmidt, M.E. and E.A. Vandewater (2008), "Media and Attention, Cognition, and School Achievemen", *Children and Electronic Media,* Vol. 18(1).

Schwartz, L. (2005), "New technologies and information overload", The Left Atrium, University Health Network, University of Toronto, Canada, *www.cmaj.ca/content/173/11/1366.*

Selwyn, N. (2009), "Faceworking: exploring students' education-related use of Facebook", *Learning, Media and Technology,* Vol. 34(2), pp. 157-174.

Selwyn, N. (2009), "Faceworking: exploring students' education-related use of Facebook", *Learning, Media and Technology,* Vol. 34(2), pp. 157-174.

Shaffer, D.W. (2008), *How Computer Games Help Children Learn*, Palgrave MacMillan, New York.

Shaffer, D.W. and K.A. Clinton (2006), "Toolforthoughts: Reexamining thinking in the digital age", *Mind, Culture and Activity*, Vol. 13(4), pp. 283-300.

Sonck, N. *et al.* (2011), *Digital Literacy and Safety Skills*, LSE EU Kids Online, London.

Stelzner, M. (2010), "*2010 Social Media Marketing Industry Report: How Marketers are Using Social Media to Grow their Businesses"*, *www.SocialMediaExaminer.com.*

Yen, C. F., *et al.* (2009*), "*Multi-dimensional discriminative factors for Internet addiction among adolescents regarding gender and age", *Psychiatry and Clinical Neurosciences.*

Young, K. (1999), "Internet Addiction: Symptoms, Evaluation and Treatment", *Innovations in Clinical Practice* (Vol. 17) in L. VandeCreek and T. L. Jackson (eds.), Professional Resource Press, Sarasota, FL.

Zickuhr, K. (2011), "Pew Internet around the web", Pew Internet and American Life Project.

Chapter 7

Key findings

There is evidence to support the notion that in OECD countries a large majority of young people, starting at an increasingly earlier age, already benefit from connectedness, that is, that they are able to use the opportunities offered by digital media and connectivity to their own advantage. Yet, when it comes to young people's expectations about technology use in learning, the resulting picture is complex. The evidence shows that young people's expectations and behaviours as learners in relation to technology use or connectivity in formal education are not changing dramatically. The vast literature defending the idea that formal education should radically change in order to cope with the expectations of young people is not supported by the facts. Empirical research has demonstrated that learners are not always comfortable with innovative uses of technology in formal education despite their social practices outside of the boundaries of educational institutions. Their attitudes stem from their prior experience in formal education, and their expectations can be succinctly reduced to three points: they expect technology to be a source of engagement, to make school or academic work more convenient, and, certainly, to make them much more productive. Yet, educators and policy makers should look at young people's current practices as a source of inspiration. Schools should not be expected to simply mimic young people's practices with technology, but this does not mean that they cannot learn from these practices and find inspiration in them. Moreover, the unprecedented challenges posed by connectedness require educators to pay attention to learners' voices.

Despite the claim by certain analysts that the recent emergence and adoption of digital technologies is no different to similar experiences in the past, and that there is therefore no reason to be too concerned about their educational implications, it does appear that the opportunities afforded by the current wave of technologies are indeed different in many respects. Contrary to what happened previously to older generations when radio and, particularly, television emerged, digital technologies and the services associated with them have brought with them something completely new: they modify not only the speed at which people deal with and manage information but also how they eventually transform it into knowledge. This is a good starting point for considering the implications that this fact may have when the users are children or young people, particularly as access to digital technologies is becoming almost universal in OECD countries.The following paragraphs summarise the main conclusions of the NML project discussed in this report:

1. **The knowledge economy and society are permeated and supported by connectedness and technology.** This has important implications for education because it has to deal with new challenges related to labour market requirements and social change. Firstly, education has to equip younger generations with the range of skills that are now demanded by the labour market in a knowledge economy. This is still a challenge in many OECD countries, particularly in relation to the development of 21st century skills. Secondly, the role that connectedness plays in new forms of socialisation and social interactions has crucial effects for the process of identity formation in adolescence. In both cases, formal education institutions have to design the best strategies to cope with these challenges since economic growth and social cohesion may be at stake.

2. **There is evidence to support the notion that in OECD countries a large majority of young people, starting at an increasingly early age, already benefit from connectedness**. Younger people do have a greater range of digital technologies in their household, tend to use the Internet as a first port of call, have higher levels of Internet self-efficacy, multi-task more and use the Internet for fact checking and formal learning activities. Nevertheless, generation was not the only significant variable in explaining these activities: gender, education, experience and breadth of use also play a part. Indeed, in all cases immersion in a digital environment (*i.e.* the breadth of activities that people carry out online) tends to be the most important variable in predicting if someone is a digital native in the way they interact with the technology. What is very clear is that it is not helpful to define digital natives and immigrants as two distinct, dichotomous generations. While there were differences in how generations engaged with the Internet, there were similarities across generations as well, mainly based on how much experience people have with

using technologies. In addition, individuals' Internet use lies along a continuum of engagement instead of being a dichotomous divide between users and non-users (Helsper and Eynon, 2010; Van Dijk, 2005; Warschauer, 2002). Clearly, connectedness suits young people's needs in domains that are critical for them such as entertainment and socialisation with peers, extending in time and intensity the influence that the peer group may have. The high levels of connectedness exhibited by the younger generations are yet an additional challenge for education. Both parents and educators should pay attention to this as well to other emerging concerns raised by connectedness and for which they lack clear guidelines based on previous experiences.

3. **Being more connected is not necessarily always a good thing: what matters is what young people do while they are connected.** Just because young people do more of something it is not always a good thing. While a strictly dichotomous classification of the effects of technology on learners into "good" and "bad" may make for nice headlines, such a simple scheme ignores the fact that human experience is intrinsically multidimensional; almost all experiences are "good" in some ways and "bad" in others. Not surprisingly, then, technology has been linked with both positive and negative effects (Bavelier *et al.*, 2010). While it is important to understand what young people are using new technologies for in debates about future developments in pedagogy and curriculum; it cannot be assumed that increased use of digital technology has a positive effect. For example, it is well known that young people multi-task more. However, we do not know if this is a positive or negative aspect of young people's use of new technology. Multi-tasking may have a negative impact on learning due to cognitive overload (Hembrooke and Gay, 2003). Similarly, while young people are more likely to use the Internet as a first port of call for information this does not mean they are in fact skilled in dealing with and critically assessing information (Livingstone, Ólafsson and Staksrud, 2011).

4. **As of today, there is not enough research evidence to demonstrate that technology attachment or connectivity has critical effects on cognitive skills development.** It may be too early to perceive significant effects. However, there are some indications that in the long run, due to continued practice, verbal intelligence levels may decrease to the benefit of image or spatial intelligence. Yet, claims about changes in the brain caused by attachment to technology or connectedness are simply not backed by evidence.

5. **The evidence shows that young people's expectations and behaviours as learners in relation to technology use or connectivity in formal education are not changing dramatically.** The vast literature defending

the idea that formal education should radically change in order to cope with the expectations of young people is not supported by the facts. Empirical research has demonstrated that learners are not always comfortable with innovative uses of technology in formal education despite their social practices outside of the boundaries of educational institutions. Their attitudes stem from their prior experience in formal education, and their expectations can be succinctly reduced to three points: they expect technology to be a source of engagement, to make school or academic work more convenient, and, certainly, to make them much more productive. At this point, some important lessons emerge:

- **Students want technology to improve teaching and learning, not to change it radically.** They value technology adoption in teaching and learning provided that it improves convenience and productivity in their academic work and school-related tasks. Teachers' perception of students' expectations regarding learning tend to overestimate students' degree of attachment to course adoption of technology. In this respect, the image of the New Millennium Learners goes far beyond the reality of the expectations of today's students and there are no indications that this will change in the soon. In particular, students' attitudes towards technology use in teaching and learning appear to be far from what many would wish to see emerging as the dominant patterns. Rather, students tend to be more reluctant in this respect than the image of the New Millennium Learner would suggest. Most of them do not want technology to bring a radical transformation in teaching and learning but would like to benefit more from their added convenience and increased productivity gains in academic work. The reasons for such reluctance might be related to the uncertainty, disruptiveness and discomfort that discrete technology-based innovations may cause for them. They may also be related to the fact that many of these students have not really experienced innovative uses of technology in their classroom.

- **Adults, specifically teachers, can 'speak the same language' as their students if they want to**. (Helsper and Eynon, 2010) – Recent evidence suggests that it is possible for adults to show the typical behaviour of digital natives, especially in the area of learning, by acquiring skills and experience in interacting with information and communication technologies (Bayne and Ross, 2011). The demographics are clearly very complicated and resistant to neat generational labelling. Clearly, much literature overestimates the impact of technology on the young and underestimates its effect on older generations (Williams and Rowlands, 2008). Evidence suggests that the differences in information behaviour, at a single point in time, between young and early middle-aged students and faculty are much less significant than those between young and more mature (40- and 50-year-old) students. A much greater sense of balance is needed. Generation is only one of the predictors of advanced interaction with the Internet. Breadth of use, experience, gender and educational levels are also important, indeed in some cases more important than generational differences, in explaining the extent to which people can be defined as digital natives. The presumed gap between educators and students may not be supported by evidence, but if such a gap does exist, it is definitely possible to close it (Helsper & Eynon, 2010).

- **Educators and policy makers should look at young people's current practices as a source of inspiration**. Schools should not be expected to simply mimic young people's practices with technology, but this does not mean that they cannot learn from these practices and find inspiration in them. Connectedness is changing the way learners acquire information and elaborate knowledge. Their identities are shaped by interacting with peers in an enlarged digital landscape of opportunities, including those for learning. As previous OECD work on the New Millennium Learners has demonstrated [EDU/CERI/CD(2008)4], there is enough empirical evidence to show that young people's use of digital media aligns with well-documented principles of social learning and knowledge management. Moreover, digital media allow a style of learning that is less about consuming knowledge and more about interaction and participation. Paying attention to how young people learn, play and socialise outside the classroom may be an important source of evidence and inspiration in the effort to introduce educational innovations. But the final criterion for technology use in learning should remain a professional judgment about the most efficient way to improve the quality of the learning experience and its results, based on sound evidence about what works.

- **The unprecedented challenges posed by connectedness require educators to pay attention to learners' voices**. The whole issue of the gap between in-school and out-of-school practices related to education raises once more the need to consider, in any educational intervention, who the learners are and how they are changing. Policy makers, professional educators and parents, each at their own level, could benefit greatly from paying more attention to what learners have to say – not only about technology use but about learning in general. Research methodologies and national or institutional monitoring mechanisms can play an important role, but nothing can substitute for an open dialogue about the ways in which learning conditions could be improved. More importantly, as already seen in the context of the New Millennium Learners Project, technology can also provide excellent opportunities to empower learners' voices in the dialogue concerning what a good learning environment should look like.

6. **In education, stereotyped concepts such as the New Millennium Learners, digital natives or net generations have to be used prudently as they can be misleading**. Today's students are heavy users of digital media and tend to benefit as much as they can from connectedness; so, in this respect they can be conceptualised as a generation of New Millennium Learners. However, there is a variety of student profiles when it comes to the intensity of attachment to technology or the variety of its uses. All of them are already in educational institutions of all levels, and it would be discriminatory to develop policies based on just one of the profiles. Therefore, terms like these can be useful to describe a social phenomenon in which, on average, younger generations show higher levels of connectedness than adults. Even more, they can also be used as a resource to evoke the range of issues raised by disparities in connectedness among generations. Beyond this generic use, however, they can obscure the most important issues at stake: individual differences and needs, the range of skills required to benefit from an educational use of technology and whose decision it is to use technology in learning. In an educational context an image that is either too generic or stereotyped may cause more harm than good, discouraging genuine debate about significant issues:

   - **These stereotypes implicitly assert that all young people are the same with regard to technology, which is far from being true**. A mixed and far more complex picture exists than is often presented in most of the well-known essays on this topic emerges from the evidence. The concept of the 'digital native' is problematic, if not entirely inadequate for policy and educational discussions (Helsper and Eynon, 2010; Thornham and McFarlane, 2011) and has to be deconstructed (Thomas, 2011), if not totally abandoned (Bennet and Maton, 2011): it is a misconception that idealises and homogenises young people's skills and interests. Available evidence, albeit still scarce, suggests diversity rather than conformity. Such an image does not adequately emphasise that socio-economic status and gender still play a critical role. If these differences are not taken into account, generic policies towards technology in education that assume all learners are equally skilled at, and interested in, technology could result in wider differences in learning results, simply by amplifying the existing socio-economic gaps.

- **For the purposes of improving teaching and learning in formal education, it is the diversity of students and situations that matters most**. The implications of young people's attachment to digital technology and connectedness are likely to be better understood by establishing a better footing for discussion and expanding the empirical research effort (Bennet and Maton, 2011). This can be done, for instance, by highlighting the significant differences within cohorts of young people in terms of their preferences, skills and use of new technologies (Kennedy *et al.*, 2008; Kennedy *et al.*, 2010). As Facer and Furlong already argued a decade ago (Facer and Furlong, 2001), young people are not a 'homogeneous generation of digital children'.
- **The skills that young people develop by themselves with regard to technology do not necessarily help them to maximise their learning opportunities.** Young people are interested in technology because of the connectedness it helps them to achieve. Connectedness provides them a tool for entertainment, for extending anytime anywhere the ability to interact with peers and, eventually, for school-related tasks – but, in the latter case, quite often without the critical approach that their teachers would like to foster. Being familiar with ICT does not necessarily entail being able to use ICT in a competent way. Living in a digital environment does not reliably imply being digitally competent. This is a consistent finding across the board, not only in OECD countries but, in fact, in very diverse societies like China and South Africa (Li and Ranieri, 2010; Thinyane, 2010). Even though new generations seem to spontaneously learn to use technologies, there is not enough evidence showing that they instantly become digitally competent as, for instance, to be much more proficient in learning using the relevant digital skills. Well-designed instructional materials for developing teenagers' digital competence are highly recommended and further research on assessing digital competence and improving ICT education and media education are urgently needed. Young people still need to be educated to make the most out of connectedness. Teachers often – and incorrectly – take for granted that the familiarity of students with technology automatically makes them savvy in information and communication skills. This is evidently not the case, and plagiarism is the clearest indication of the lack of adequate education in this domain. The range of digital skills that most students possess does not easily translate, without guidance, into improved learning skills.
- **These images implicitly convey the message that learners are urging institutions and teachers to adopt technology, which,**

at least today, is an oversimplification that confuses both policy makers and educators. The latter need to be reminded that connectedness is yet another tool at their disposal and that decisions about technology adoption have to be taken in the light of professional judgment, based on evidence. For instance, there are many calls for the use of Web 2.0 technologies in classes. The rationale can be described as follows: "It's fun and cool to blog; lots of people are doing it; we know that kids get some information from blogs; therefore, blogs must have a place in our schools" (Palfrey and Gasser, 2008, p. 248). But this rationale does education no favours, and could result in exhausting teachers' efforts to keep up with technology developments. The "novelty factor" has, by definition, a short-lived nature (Glover and Miller, 2001; Saunders and Klemming, 2003) and shouldn't be used to replace a sound pedagogical foundation: the main reason for adopting a particular technology should be that it allows methodological change, promises improved results and offers greater learner satisfaction.

References

Bavelier D., *et al.* (2010*)*, "Removing brakes on adult brain plasticity: from molecular to behavioral interventions", *Journal of Neuroscience*, Vol. 30, pp.14964-14971.

Bayne, S. and J. Ross (2011), "'Digital Native' and 'Digital Immigrant' Discourses", in R. Land and S. Bayne (eds.), *Digital Difference*, Vol. 50, SensePublishers, London.

Bennet, S. and K. Maton (2011), "Intellectual Field or Faith-based Religion. Moving on from the Idea of 'Digital Natives'", in M. Thomas (ed.), *Deconstructing digital natives. Young people, technology and the new literacies*, Routledge, New York and London.

Facer, K. and R. Furlong (2001), "Beyond the myth of the 'Cyberkid': young people at the margins of the information revolution", *Journal of Youth Studies,* Vol. 4(4), pp. 451-469.

Glover, D. and D. Miller (2001), "Running with technology: The pedagogic impact of the large scale introduction of interactive whiteboards in one secondary school", *Technology, Pedagogy and Education,* Vol. 10(3), pp. 257-278.

Helsper, E. and R. Eynon (2010), "Digital natives: where is the evidence?", *British Educational Research Journal,* Vol. 36(3), pp. 502-520.

Hembrooke, H. and G. Gay (2003), "The laptop and the lecture: the effects of multitasking in learning environments", *Journal of Computing in Higher Education,* Vol. 15(1), pp. 46-64.

Kennedy, G. *et al.* (2008), "First year students' experiences with technology: Are they really digital natives?", *Australasian Journal of Educational Technology,* Vol. 24(1), pp. 108-122.

Kennedy, G. *et al.* (2010), "Beyond natives and immigrants: exploring types of net generation students", *Journal of Computer Assisted Learning,* Vol. 26(5), pp. 332-343.

Li, Y. and M. Ranieri (2010), "Are 'digital natives' really digitally competent? A study on Chinese teenagers", *British Journal of Educational Technology,* Vol. 41(6), 1029-1042.

Livingstone, S., K. Ólafsson and E. Staksrud (2011), *Social Networking, Age and Privacy*, LSE EU Kids Online, London.

Palfrey, J. and U. Gasser (2008), *Born digital: Understanding the first generation of digital natives*, Basic Books, New York.

Saunders, G. and F. Klemming (2003), "Integrating technology into a traditional learning environment", *Active Learning in Higher Education,* Vol. 4(1), pp. 74-86.

Thinyane, H. (2010), "Are digital natives a world-wide phenomenon? An investigation into South African first year students' use and experience with technology", *Computers and Education*, *55*, pp. 406-414.

Thomas, M. (ed.) (2011), *Deconstructing digital natives. Young people, technology and the new literacies*, Routledge, New York and London.

Thornham, H. and McFarlane, A. (2011), "Discourses of the Digital Native", *Information, Communication and Society,* Vol. 14(2), pp. 258-279.

Van J. Dijk (2005), *The deepening divide: inequality in the information society*, Sage, Thousand Oaks.

Warschauer, M. (2002), "Reconceptualizing the digital divide", *First Monday,* Vol. 7(7).

Williams, P. and I. Rowlands (2008), *Information behaviour of the researcher of the future: the literature on young people and their information behaviour*, British Library and JISC, London.

Chapter 8

Implications for educational policy, research and practice

The main findings of the OECD's project on the New Millennium Learners suggest that there is a need for identifying which policies and practices will best suit the objective of providing all students with a rich learning environment based on their particular profiles and needs, while improving their satisfaction and boosting learning gains. To do so, more must be done to improve the knowledge base about technology use in education so as to inform these debates. This requires not only more experimental research but also increased efforts to better disseminate existing findings and avoid reinventing the wheel. Activities intended to train and support teachers for course adoption of technology should be based on validated effective practices and take into account students' alternative profiles, needs and expectations.

Inevitably, policy makers and educators will ask questions about what the implications of these conclusions for them are, and whether some particular courses of action should be adopted or not. While educational institutions and teachers are increasingly adopting technology in teaching, there is an urgent need to address these issues in a systemic way (OECD, 2010b). The following summarises what should be the key components of the public agenda on connectedness and learners and the corresponding recommended actions for policy makers, teachers, parents and researchers.

The agenda for the public debate

One of the most visible benefits of the overall discussion about the New Millennium Learners is the fact that the whole issue of connectedness and learners is part of public debates about education worldwide. This attracts a lot of media attention, particularly when the discussions focus mostly on the negative direct or indirect effects of technology attachment. Parents and teachers may feel that they are somehow lost in this debate and look to each other in search of valuable and practical advice. Governments may think that their responsibility should be limited to the provision of the required investments to connect schools and let teachers address what to do with it themselves.

There is a need for a public debate about youth, education and connectedness with an agenda that is informed by evidence. In the light of the key findings of this report, such an agenda should include the following themes:

1. **Learners' diversity.** Learning styles and student profiles are a recurrent theme in educational research. Yet, educational debates and practices regarding technology and learning hardly take into account one of the major findings of this report: that there are different student profiles also in relation to connectedness, and that they have different educational needs and expectations when it comes to the use of technology in education.

2. **Digital literacy.** Far from being an innovative concept, digital literacy has been in the educational discussions for some time. Yet, far from being restricted to the basic skills that allow people to use computers and the Internet, digital literacy has to be revisited in light of the findings of this report. It has to embrace the wide range of skills on which basis individuals can actively participate in knowledge economies and societies. In revisiting digital literacy two areas have to be explored:

- School curriculum: in most OECD countries, courses devoted to digital literacy have disappeared from the curriculum under the assumption that children do not need any more computer literacy. Yet, there is a need to reconsider whether the approach according to which all teachers are held equally responsible for teaching how to develop critical digital skills is really working, that is, whether school education is addressing the needs of today's students in this respect.
- Teacher standards and training: similarly, teacher training institutions do not seem to be the avant-garde of digital literacy, as they would be expected to be. If student teachers lack the opportunity to become digitally literate throughout their initial training the chances that they become effective once placed in schools are left to individual engagement and exploration. There is an urgent need to re-define the teaching profession in the context of the knowledge economy, set the appropriate standards and monitor how they translate into effective practice.

3. **The new digital divides.** The work on the New Millennium Learners has been also instrumental in highlighting the emergence of new digital divides, which go beyond access to computers and the Internet. As this report shows, there are persistent gaps in access even in the most advanced OECD countries. Yet, the new digital divides are related to the ability of individuals to make significant use of connectedness and reap the corresponding benefits as well. Education has an important role to play in the struggle against these more subtle digital divides, provided that tailored policies are in place, both at government and institutional levels. In particular, at least in OECD countries, the current emphasis on policies intended to grant access to equipment, as exemplified in one-to-one computer policies, has to move towards an even greater focus on meaningful applications of connectedness in education – taking advantage of the fact that students already go to school with their own equipment, although schools may not acknowledge it.
4. **The blurring boundaries between formal and informal learning.** A close examination of how young people benefit from connectedness shows that there is a lot of informal learning happening. There is a missing opportunity if this is not consolidated with the kinds of formal learning happening in the classroom, not only because schools have difficulties in acknowledging that informal learning is, above all, learning that can be meaningful but also because teachers do not know how to bridge the gap between formal and informal learning. Some studies argue that the impact of home interaction with the

Internet on formal education should not be ignored because they influence what young people are able and willing to learn in school. Questions that educators need to ask themselves are, for example, "are we only interested in supporting formal learning activities or are we satisfied if people just focus on playing online games, further developing skills they acquired in informal contexts?" (Helsper and Eynon, 2010). Connectedness may offer the bridging opportunities that would make learning move seamlessly between the classroom and the world outside, already populated with social applications where information and knowledge are currently being shared by students. Teachers need to acknowledge this and use it for the benefit of consolidated learning.

5. **Technology for monitoring and assessing learning.** There is no doubt about the benefits of using technology for the administration of student assessments, as well as for monitoring learning over time, in comparison to traditional means. Learning management systems have proved to be worthwhile in this respect. The new frontier is, however, how to make the most out of technology to assess the complex, second order skills that are required by the knowledge economy and often supported or at least enhanced by technology. Technology is currently able to provide the type of solutions that address this assessment need. Clearly, the use of these technology-supported solutions would have the added benefit of a cascade effect in the classroom.

6. **Monitoring changes in learners' needs and demands.** When it comes to connectedness, the future may bring even more impressive changes than those already accounted for in this report. Traditional education systems have had difficulties in paying attention to how learners' needs and demands are evolving following societal and technological change. Yet, for educational systems it should be imperative to design systems and methodologies, such as student surveys, by which to better understand the transformation of learners' needs and expectations. More importantly yet is the ability to disseminate the knowledge obtained to teachers and parents in a meaningful way.

Each of the aforementioned themes requires responses from the key stakeholders – policy makers, teachers and parents – and poses new challenges to researchers in education. What follows is an account of the different implications of the key findings of this report for each group of stakeholders.

Implications for policy makers

Governments have done a lot to support technology adoption in teaching, with important investments in infrastructure as well as in services both for students and teachers, and educational institutions and their networks as well. Governments must keep up with emerging technology developments, equipment and applications, and contribute to supporting innovations intended to explore the value and possible benefits of technology adoption for teaching and learning.

On top of this, the whole issue of the New Millennium Learners, when duly revisited, urges some policy developments in the following areas:

- Although the need for the development of 21st century skills seems to have permeated policy discourse, there is a pressing need to see how this translates into curriculum reform, teacher training and professional development and, more importantly, assessments; without changes in these policy areas educational practices will remain unchanged.
- Given uneven digital skills and the policy and educational discussions about the need to identify more precisely the kinds of skills required, an international inventory and agreed framework for digital literacy would provide a valuable resource for policy makers, teachers, parents and the industry.
- Contrary to how it seems at first glance, research has demonstrated that learners' connectedness does not automatically translate into a capacity to use it to maximise learning opportunities; if schools and teachers have to assist learners in doing so, the necessary capacity has to be built in schools by providing appropriate technical conditions, and more importantly, institutional and professional incentives for innovation in this domain, dissemination and scaling up;
- Distributive policies intended merely to universalise access to devices (be they interactive whiteboards or personal computers on a one-to-one basis), although contributing to the realisation of the vision of a more suitable education for the knowledge economy and society and better fitted to the stereotype of a connected generation, should be closely monitored and assessed in relation to policies' ability to bring change into the classroom and improve the learning experience and outcomes of learners; otherwise, such policies risk becoming a very costly commitment with few returns.
- Moreover, there are subtle forms of digital divides that need to be addressed with tailored policies that have explicit compensatory goals; distributive policies (as mentioned above) may unintentionally

contribute to enlarging the impact of socio-economic status on learning outcomes if not duly accompanied with substantive policies addressed to monitor change and effects on learning outcomes. For children who still lack access, efforts are vital to ensure digital exclusion does not compound social exclusion. As the pace of change in this domain accelerates, there is a need for mechanisms that can help us to understand the challenges ahead; it is important to look at the evidence concerning what technologies students have access to and what their preferences are in order to inform both policy and practice; this is particularly appropriate when it comes to the educational potential of emerging technologies – as is likely to be the case with 3D. Monitoring systems intended to survey learners' practices with digital media and connectedness and national forums designed to help learners raise their voices can also be helpful.

Implications for educational institutions and teachers

Because of the growing importance of connectedness, schools and teachers must cope with new responsibilities related in particular to skills with which they may not be as familiar as necessary. If this is the case, they should find the appropriate means to develop the capacities required both professionally and institutionally.

Schools play a pivotal role in digital skills development, mitigating forms of digital exclusion. Inequalities in digital skills persist – in terms of SES, age and, to a lesser degree, gender, so efforts to overcome these are needed (Sonck et al., 2011). Schools should take a major responsibility for supporting children and their parents in gaining digital literacy and safety skills. Such efforts should become established as a core dimension of the curriculum, and initiatives developed at secondary school level should now be extended to primary and even early childhood education. Digital skills for all ages remain important but younger age groups need to be a particular priority for parents and teachers (O'Neill, Livingstone and McLaughlin, 2011). Although it is commonly said that teachers are often inadequately resourced and trained to carry out the functions entrusted to them, an increasing percentage of them can hardly be considered digital immigrants. The adoption of technology has contributed to transforming teachers' work although this process is slower in the schools sector than in higher education. There are indications that the actual use of technology in teaching in higher education clearly outperforms the equivalent in the schools sector in most OECD countries. The gap in technology adoption between students and teachers is decreasing although the range of applications and services used by them differ. The key message here is that neither schools nor teachers can be said to be closing their eyes to changes in students' behaviours, needs and expectations. But the responsiveness of education systems to them could be quicker.

Evidence from OECD studies in initial teacher training [EDU/CERI/CD(2009)21] also suggests that the issue of technology use is not addressed adequately and that student teachers are not receiving enough training or practical experience in how they can use technology in innovative ways that can improve or transform their classroom practice. This unfortunately implies that a new generation of teachers, many of whom are New Millennium Learners themselves, are starting their careers ill-prepared in terms of the potential pedagogical benefits of new technologies.

Yet teachers experience difficulties applying these principles in traditional learning environments – partly, but not exclusively, because the intensity of technology use in classrooms does not seem to have reached a critical threshold to produce significant results. PISA evidence (OECD, 2010a) suggests that there are clear benefits for educational performance associated with technology use, provided that a critical threshold of use is reached. Today's estimated use of technology in compulsory education in EU countries (Empirica, 2006) is less than one hour weekly, which compares poorly with the daily experience of those same students at home. Yet, educational institutions, teachers, parents and policy makers still seem to be waiting for definitive evidence and advice on how to deal with these connected minds. In fact, there seems to be a problem of failure of systemic innovation (OECD, 2010b) in the particular domain of technology use in formal education. This failure urges additional implementation research: what the results of different policies, strategies and plans to foster technology use in the classroom are remains an unaddressed research question from an international and comparative perspective.

As of today, the most pressing challenges for schools and teachers are in the following areas:

- How to progressively integrate the new digital media and the resulting innovative social practices into the daily experience of schooling.
- How to integrate the teaching of 21st century skills, particularly those supported with or enhanced by connectedness, with subject-related skills and, more specifically, how to assess them.
- How to help learners to develop, in particular, the digital and information skills that it is so often assumed, mistakenly, that the young generations possess by nature. Strengthening information literacy is vital in order to improve the educational use of connectedness.

- How to specifically address the development of digital skills that could help learners to make the most of connectedness for a more enjoyable, convenient and productive learning experience.
- How to deal with all these emerging needs while at the same time paying attention to the differential needs of learners and providing increased support to those who come from disadvantaged socio-economic contexts.

Implications for parents

Irrespective of the lack of universal validity for the concept of the New Millennium Learners, parents should consider what parenting means in the age of connectedness. As a matter of fact, parents have to be reminded that they have to be involved in all spheres of the development of their children – and the role that digital media and connectedness are increasingly playing in children's lives makes it inevitable that parents will need to assume responsibility in this sphere. Many parents appear to have got the message that it is valuable for them to engage with their child's Internet use, and they employ a wide range of strategies, depending partly on the age of the child. But there are some parents who do not do very much, even for young children, and there are some children who wish their parents to take more interest. Targeting these parents with awareness raising messages and resources is thus a priority.

Particular issues that require their attention include education for a safe use of connectedness and the need to help their children to balance the time devoted to online activities, including extended interaction with peers, with other activities. Parents should be reminded that if their children lack support in this important part of their lives, there is a risk that they will grow up as digital orphans – without any parental guidance or support to maximize the opportunities offered by connectedness and cope with the risks. Cynicism that what parents do is not valued, or that children will evade parental guidance, is ungrounded: the evidence (O'Neill, Livingstone and McLaughlin, 2011) reveals a more positive picture in which children welcome parental interest and mediating activities while parents express confidence in their children's abilities. It is important to maintain this situation as the Internet becomes more complex and more embedded in everyday life.

Those parents lacking the appropriate conditions for dealing with connectedness will inevitably turn to schools and teachers in search of answers and support. Governments should take this into account and provide dedicated channels to cope with the information and capacity needs of families with regard to connectedness, without overloading schools.

Implications for research

Since the advent of personal computers there has been a dramatic increase in the amount of technology available to and used by children and young people. This is bound to shape the way they learn, develop, and behave. Given the multifaceted nature of technology, it is perhaps unsurprising that the story of its impact on learners is extremely complex and multisided. Some forms of technology have no effect on the form of behavior they were designed to transform, while others have effects that reach far beyond their intended outcomes. All of this is indicative of a field that is still emerging. What we do know is that, in technology, we have a set of tools that has the capability to drastically modify human behavior. What remains, which is not trivial, is to determine how to purposefully direct this capability to produce desired outcomes. In this endeavor it is imperative to enrich the educational debates with behavioral and neuroscience research (Meltzoff *et al.*, 2009).

There is a need to invest in empirical research in order to elucidate in which ways technology can provide more than convenience and productivity – in particular, learning benefits – either by providing a more rewarding experience or better learning outcomes, or both at the same time. As Dede (2007, p. 4) has already outlined, one starting point for fruitfully locating technology in higher education pedagogy is to observe how students are using technology in other aspects of their lives, "sifting out the dross of behaviours adopted just because they are novel and stylish from the ore of transformational approaches to creating, sharing, and mastering knowledge." There can be a tendency within educational policy to see technology as the 'fix' or 'solution' to many of the challenges the sector faces and there is a danger that the current popularity of statements about young 'techy' generations could increase the prominence of this deterministic view. To counter such claims, the publication and discussion of empirical work on the realities of how younger and older generations learn through, and engage with, technology is needed. What is at least as important as the research effort is the ability to share the results in forums where they can be translated into recommendations for better practice. This should not be an individual task, but a commitment of the whole educational community.

The four main challenges for research in relation to the educational implications of connectedness are the following:

- Supporting research about the educational potential of emerging technologies.
- Continuing to monitor the effects of connectedness on learners, supplementing surveys with observational activities particularly in relation to social applications, thus supplying the best available evidence to understand technology-induced changes in learners.

- Improving, through experimental research, the understanding of the conditions under which technology adoption can result in a more enjoyable, convenient and productive experience of learning.
- Providing policy makers and practitioners with empirically supported evidence about what works in this domain and how to scale it up because decisions regarding the use of technologies for learning should not only be based around students' preferences and current practices, even if properly evidenced, but on a deep understanding of what the educational value of these technologies is and how they improve the process and the outcomes of learning.

Concluding remarks

Some well-established facts and research findings have to be considered a consistent foundation for action. To begin with, empirical research has demonstrated that, with a change in the way teachers teach, students can have better results regarding higher-order thinking and 21st century skills (Langworthy, Shear and Means, 2010). Additionally, there is a clear correlation between a more intense use of digital media and educational performance (OECD, 2010a). Yet, educational research regarding the implementation of technology in education has not been systematic enough (OECD, 2010b). All things considered, it is clear that there is a large unused potential to boost students' results and to improve educational efficiency or productivity by:

i) more cumulative research on how to implement technology in education to support methodological change;

ii) systematically changing the way teachers teach; and

iii) keeping educational institutions in line with contemporary technology developments that have an impact on human communication, socialisation and knowledge management.

A couple of questions emerge inevitably: What will the future bring? How should education prepare for that future?

If anything is clear it is that technology will continue to evolve as fast as it has done in the past decade, if not faster. Digital devices that are considered to be indispensable by today's students were not accessible to a majority of them even five years ago if not less. The future will bring also new applications and environments (Johnson *et al.*, 2011) that once again may have an impact on the way young people communicate, are entertained, socialise, and deal with their coursework.

It is unclear, however, if the new technology developments will transform students' learning expectations and demands. Drawing on the past years, a prudent approach would be to state that a certain evolution will take place, particularly if the experiences with technology in the previous school years contribute to raising students' awareness of the advantages of using technology for learning. Peers teach how to make the most out of connectedness for nurturing social relationships that matter. Schools should draw on that to teach students how to benefit from connectedness in a way which results in better learning and improved outcomes. Parents should educate their children to become responsible individuals in a context where connectedness poses new challenges, full of both threats and opportunities.

This report has shown that the New Millennium Learners is a broad expression that, rather than describing a homogeneous generational phenomenon should be taken as a point of departure for a public debate about education in the digital age. All concerned stakeholders, from policy makers to researchers, from teachers to parents, and whenever appropriate also learners themselves, should do their best to express their views, expectations, needs and concerns in a way that can easily inspire public interventions, teachers' professional judgment and action, and parenting.

No one can predict what the teaching and learning experience in education will be like in a decade. The recent evolution shows that what works in relation to technology in education comes as a result of the dialogue between students who master digital media but have quite prudent expectations about their use in teaching, and teachers who want to extend the benefits of convenience and productivity brought about by technology to enrich their teaching responsibility. It is in the best interest of institutions to nurture this ongoing dialogue with accompanying measures and incentives and to let students express their own voices. Such a dialogue should remain as open as the future is.

References

Dede, C. (2007), Foreword, in G. Salaway, J.B. Caruso and M.R. Nelson (eds.), *The ECAR Study of Undergraduate Students and Information Technology, 2007*, Vol. 6, Educause, Boulder, CO.

Empirica (2006), *Benchmarking Access and Use of ICT in European Schools 2006 – Results from Head Teacher and Classroom Teacher Surveys in 27 European Countries*, European Commission, Brussels.

Helsper, E. and R. Eynon (2010), "Digital natives: where is the evidence?", *British Educational Research Journal*, Vol. 36(3), pp. 502-520.

Johnson, L. *et al.* (2011), *The 2011 Horizon Report*, The New Media Consortium, Austin, Texas.

Langworthy, M., L. Shear and B. Means (2010), "The third lever: Innovative teaching and learning research", OECD (2010b), *Inspired by Technology, Driven by Pedagogy: A Systemic Approach to Technology-Based School Innovations*, OECD Publishing, Paris.

Meltzoff, A.N. *et al.* (2009), "Foundations for a new science of learning", *Science,* Vol. 325, no. 5938, pp. 284-288, *www.sciencemag.org/content/325/5938/284.*

O'Neill, B., S. Livingstone and S. McLaughlin (2011), *Kids Online: Final recommendations for policy, methodology and research*, LSE EU Kids Online, London.

Organisation for Economic Co-ordination and Development (OECD) (2010a), *Are the New Millennium Learners Making the Grade? Technology Use and Educational Performance in PISA*, OECD Publishing, Paris.

OECD (2010b), *Inspired by technology, driven by pedagogy: A Systemic Approach to Technology-Based School Innovations*, OECD Publishing, Paris.

Sonck, N. *et al.* (2011), *Digital Literacy and Safety Skills*, LSE EU Kids Online, London.

ORGANISATION FOR ECONOMIC CO-OPERATION AND DEVELOPMENT

The OECD is a unique forum where governments work together to address the economic, social and environmental challenges of globalisation. The OECD is also at the forefront of efforts to understand and to help governments respond to new developments and concerns, such as corporate governance, the information economy and the challenges of an ageing population. The Organisation provides a setting where governments can compare policy experiences, seek answers to common problems, identify good practice and work to co-ordinate domestic and international policies.

The OECD member countries are: Australia, Austria, Belgium, Canada, Chile, the Czech Republic, Denmark, Estonia, Finland, France, Germany, Greece, Hungary, Iceland, Ireland, Israel, Italy, Japan, Korea, Luxembourg, Mexico, the Netherlands, New Zealand, Norway, Poland, Portugal, the Slovak Republic, Slovenia, Spain, Sweden, Switzerland, Turkey, the United Kingdom and the United States. The European Union takes part in the work of the OECD.

OECD Publishing disseminates widely the results of the Organisation's statistics gathering and research on economic, social and environmental issues, as well as the conventions, guidelines and standards agreed by its members.

OECD PUBLISHING, 2, rue André-Pascal, 75775 PARIS CEDEX 16
(96 2011 03 1 P) ISBN 978-92-64-07522-1 – No. 60141 2012-05